The Resurrection of Man

The Story of Adam and Eve Told Mathematically

First Edition

The Resurrection of Man (The Theory of Creation): The Story of Adam and Eve Told Mathematically

© Copyright 2020 by The Heart of Man Publications, LLC

Copyright Office Registration Number: TXu-2-215-134

ISBN: 978-1-7368598-0-3

Until and unless notified, copies of the material found herein are authorized only by students, instructors, and others solely for personal use. To include any original material or original presentation of material contained herein in a publication requires prior written approval from The Heart of Man Publications, LLC.

All Rights Reserved. No part of this book may be reproduced, stored in a retrieval system, transmitted, or translated into machine language, in any form or by any means, electronic, mechanical, photocopying, recording, or otherwise, without the prior written permission of The Heart of Man Publications, LLC, except for brief quotations embodied in research articles and reviews. Any permissible public use must cite this book as a source.

Bible Version. American Standard Version of the Old Testament [281]. Book-Chapter.Verse identifies Bible verses. For instance, 1-1.26 references Genesis-Chapter 1.Verse 26.

Table of Contents

Messages to the Reader iv
Preface v
The Beginnings viii
Chapter 1 1
God's Logic 1
1.1 The Syntax of God's Logic 1
   1.1.1 God's Language 1
   1.1.2 Statements of Humans about Themselves 2
   1.1.3 Humans 2
      1.1.3.1 To Thine Own Self be True 3
      1.1.3.2 Human Deception 3
      1.1.3.3 Sense of Completeness and Incompleteness 4
1.2 Joy and Sadness 5
   1.2.1 Joy 5
   1.2.2 Sadness 5
   1.2.3 Vacillation between Joy and Sadness 5
1.3 Human Relationships 6
   1.3.1 Good and Bad Relationships 6
   1.3.2 Relationships between Collections of Humans 7
   1.3.3 Hierarchical Relationships 7
      1.3.3.1 Similarity of Two Hierarchies 7
1.4 Human Hierarchies 7
1.5 Human Paradoxes 8
1.6 Concluding Remarks 8
Chapter 2 12
God's Numbers 12
2.1 Natural Numbers 12

2.2 Integers 13
2.3 Rational Numbers 15
2.4 Real Numbers 16
2.5 Complex Numbers 16
    2.5.1 Number Entanglement...17
        2.5.1.1 Complex Number Systems 19
2.6 Concluding Remarks 20
Chapter 3 21
How God Made the Universe 21
3.1 God's Functions..22
3.2 God..23
    3.2.1 The Four Functions of God...25
    3.2.2 The Birth of Man 25
    3.2.3 Man's View 29
    3.2.4 Man's Place 30
    3.2.5 Satan 33
3.3 Concluding Remarks 34
Chapter 4 35
How God Made Space-Time 35
4.1 The Principle of Relativity 35
    4.1.2 Lorentz Transformations 37
4.2 Time 37
4.3 Concluding Remarks 40
Chapter 5 41
Epilogue...41
The Nature of the Universe 41
5.1 Small-Scale Physics 41
5.2 Large-Scale Physics..41

5.3 Combining the Small with the Large 42
5.4 The Wave Function 44
5.5 Final Remarks 45
Appendices 47
List of References 50

Messages to the Reader

Inspiration for this book comes from unconventional academic research conducted by independent investigators. Although the research does not originate from Scripture, the story that emerges bears a striking resemblance to Genesis 1, 2, and 3, which inspired spreading the word to a broader audience. No attempt is made to appeal to specific sectors of the Christian, Jewish, or Islamic faiths. The significance of the words written herein is left to the judgment of the reader.

— The Heart of Man

The story is told mathematically, mainly in "set-theoretical" language. The reader should review "The Beginnings" (page viii) before starting the first chapter. A brief description there explains the mathematical symbolizing utilized in the book. Hopefully, familiarizing the reader with the book's symbols will achieve a more enjoyable reading experience.

— The Heart of Man

Preface

"There exists, if I am not mistaken, an entire world which is the totality of mathematical truths, to which we have access only with our mind, just as a world of physical reality exists, the one like the other independent of ourselves, both of divine creation."

— *Charles Hermite*

The theory of general relativity, published in 1915, and the standard model of particle physics, completed by about 1975, are, today, the predominant theoretical systems that describe the nature of "our universe." But each system is deficient in some respects, and they conflict with one another. A flurry of active research seeks an approach that would resolve the conflicts and deficiencies associated with the present theories. Currently, "string theory" is the dominant approach, as it strives to surpass the present-day understanding of "our universe." It replaces the dimensionless point particles of the standard model with one-dimensional "strings." What was once an ill-defined "particle" is, instead, a vibrating string. Each "mode of vibration" of the string represents a different particle. From this beginning, general relativity and the standard model's gauge theories, which describe nuclear interactions, arise in a surprising fashion [12]. String theory foretells a *"theory of everything."*

But string theory lacks clarity. There is not one but five superstring theories, none of which describe "our universe." Each theory comes with six extra spatial dimensions in addition to the four dimensions of relativity theory and the standard model. The only feasible way of including the extra dimensions is to curl them into tiny compactifications. But the curling up of the extra dimensions, achieved in innumerable ways, results in thousands of string theory variations. None of them appear compatible with the standard model [14].

A conjecture presented in 1995 theorized that the five superstring theories were part of a more fundamental theory named "*M*-theory" [37]. But *M*-theory, incompletely defined, attracts criticism for providing scantly few "theoretical predictions" and being unfinished and untestable. In addition to the strings, it includes higher-dimensional objects called "branes."

Branes are the entities to which the ends of the strings can attach themselves. While enhancing the flexibility of string theory, the introduction of the "brane" increases the number of possible string theories to 10^{500} or more [17]. Whether any of the many string theory variations describe "our universe" remains an open question.

Critics argue that string theory has a mythical character, makes no testable predictions, and is unscientific. Some theorists endorse a "landscape" of string theories. Each string theory in the landscape represents a possible universe or part of a universe with a unique set of physical laws. Such a landscape, constrained by the anthropic principle, suggests that the theory which describes "our universe" must contain parameters that allow for human existence. Only the string theories with finely tuned parameters, enabling human observers to evolve, merit consideration as a model for "our universe." The anthropic principle is controversial. In its stronger form, it often serves as an argument that a God of the universe exists or, in its weaker form, as an argument against God's existence [13,15,16].

There are additional contenders for a theory of "our universe." For instance, "loop quantum gravity" asserts that space is not continuous but broken into tiny chunks connected by links. The links, tangled into braids and knots, represent elementary particles. But its proponents have struggled to incorporate a concept of "gravity" into their theories.

There are less well-known contenders for a theory of "our universe." But the real "*theory of everything*" lies in the story of Adam and Eve told mathematically, explained herein. God created the universe by numbers.

Out of logic emerges mathematics. But are the rules of logic invented or discovered? Are numbers created by God or Man? If God created numbers, then Man must find them. Some will say Man "invents" numbers. Others will declare Man "discovers" them. For instance, the fundamental idea within mathematics is the "set," designated "$S = \{m\}$." The "set" is a collection of distinct elements "m" into a whole "S." The elements are usually numbers but can be anything arising from thought or intuition. But is the "set" a discovery or an invention? Most practitioners regard the "set" as an invention because it stands alone. But not everyone agrees.

Practitioners, who believe that the rules of logic are invented, are commonly called "logical nominalists." "Logical realists" are practitioners who believe the rules of logic are discovered. Today, mathematics is highly nominalistic because virtually all mathematics arises formally from the concept of a "set," a body of work called "axiomatic set theory." Fundamentally, all mathematical systems start with a set of statements called "axioms," assumed to hold throughout the system. The remainder of the statements, the "theorems," follow directly from the axioms. But not everyone agrees that a "set of axioms" is an invention. Some people think axioms are discovered. So, who is right?

Nominalistic logic is singular because the idea of a "set" originates in the brain, the place where effectively all mathematics is invented. But the realist believes in an independent incorporeal mind, where mathematical forms reside, forms that are entirely missing from modern logic. The celestial mind is where mathematical objects are discovered. If the brain invents a set "$S = \{m\}$," S emanates from the brain. Left out of the current formal rules of logic is a discovery set "$\bar{S}$," which resides in an independent mind. But the logic chronicled in Genesis requires this. And therein lies Man's story.

The Beginnings

Use of Symbols		
Description	**Symbol**	**Remark**
An element of a "set"	Lowercase letters, "m"	"m" is a distinct element.
Set	Uppercase letters, "$\{m\}$"	$S = \{m\}$: "S" is a collection of distinct elements "m" into a whole.
Subset	$\subset$	$A \subset B$: Every element in "A" is also in "B."
Set inclusion, exclusion	$\in, \notin$	$s \in S$: "s" is an element of "S"; $s \notin S$: "s" is not an element of "S."
Universal quantifier	$\forall$	$\forall a$: A property that applies to all the elements in a set.
Existential quantifier	$\exists$	$\exists a$: A property that applies to at least one element in a set.
Uniqueness	$!$	$!a$: "a" is a unique element.
Logical conjunction	$\wedge$	$a \wedge b$: Both "a" and "b" must be true.
Logical disjunction	$\vee$	$a \vee b$: At least one of "a" or "b" must be true.
Logical negation	$\sim$	$\sim a$: If "a" is true, then "$\sim a$" is false.
Number of elements in a "set"	$\#$	$\#S$: The number of elements in "S."
Scientific notation	10^{number}, "$c \times 10^b$"	10^b: Ten to the power of $b \geq 0$; $10^{-b} = 1/10^b$.
Equal, not equal	$=, \neq$	$a = b$: "a" is equal to "b"; $a \neq b$: "a" is not equal to "b."
Greater than, less than	$>, <$	$a > b$: "a" is greater than "b"; $a < b$: "a" is less than "b."

Use of Symbols (Continued)

Description	Symbol	Remark
Relationship between the elements of two "sets"	$\xrightarrow{r}$, $r(s) = \bar{s}$	$S \xrightarrow{r} \bar{S}$: The elements of "$S$" are related to the elements of "$\bar{S}$" through a relationship "r."
Ordered pair	$\langle , \rangle$	$\langle x, y \rangle$: "x" is the first element, and "y" is the second element.
Cross product of two sets	$\times$	$\forall x \forall y (\langle x, y \rangle \in A \times B)$, $x \in A, y \in B$: "$\langle x, y \rangle$" are all the ordered pairs, where $x \in A$ is the first element, and $y \in B$ is the second element.
Relations	"$R = (A, B, C)$"	If $x \in A$, $y \in B$ and $\langle x, y \rangle \in C$, then $C \subset A \times B$.
Precedence, successor	$\preccurlyeq, \succcurlyeq$	$a \preccurlyeq b$: "a" precedes "b," $a \succcurlyeq b$: "a" is the successor of "b."
Square root	$\sqrt{number}$	If $b^2 = a$, then "$\pm b$" are the square roots of "a."
Function	"$f(t)$"	$f(t)$: "$\langle t, f \rangle$" is a relation such that for each "t," "f" has one and only one value.
Empty set	$\emptyset$	$S = \emptyset$: "S" is the "set" with no elements in it.
Comparable, not comparable	$\parallel, \nparallel$	$a \parallel b$: "a" is comparable with "b"; $a \nparallel b$: "a" is not comparable with "b."
Adding up of several numbers	$\int f(t)$, $\sum f(t)$	Adding up all or some of the numbers that "f" produces.
Logical implication	$\rightarrow$	$a \rightarrow b$: if "a" is true, "b" cannot be false.
Absolute value	$\|number\|$	$\|a\| = a$, $a \geq 0$ and $\|-a\| = a$.
Factorial	$number!$	$n!: n(n-1)(n-2) \cdots 1$.

Use of Symbols (Continued)		
Description	Symbol	Remark
Approximately equal to	$\approx$	$a \approx b$: "a" approximately equals "b."
To the power of	$number^{number}$	a^b: "a" multiplied by itself "b" times.
Never-ending	∞	Infinity.

Hierarchial Relationships		
Description	Statement	Remark
Partial and total hierarchies $R = (A, A, C)$	1. $\forall a(a \in A \rightarrow a \leqslant a \in C)$ 2. $\forall a \forall b(a, b \in A \land a \leqslant b \in C \land b \leqslant a \in C \rightarrow a = b)$ 3. $\forall a \forall b \forall c(a, b, c \in A \land a \leqslant b \in C \land b \leqslant c \in C \rightarrow a \leqslant c \in C)$	If R is a partial hierarchy on A, then so is R^{-1}, where the precedence reverses, i.e., $a \geqslant b$. If $a \leqslant b \in C$ and $a \leqslant c \in C$, but $b \leqslant c \notin C$, then "b" and "c" are "not comparable," signified "$b \not\Vert c$." A partial hierarchy "R", where all the elements of A are comparable, is a "total hierarchy."
First element	$\forall x \exists ! a(a, x \in A \rightarrow a \leqslant x \in C)$	The first element "a" is unique.
Last element	$\forall x \exists ! a(a, x \in A \rightarrow a \geqslant x \in C)$	The last element "a" is unique.
Minimal element	$\forall x \exists a(a, x \in A \rightarrow a \leqslant x \in C)$	A minimal element may not be unique.
Maximal element	$\forall x \exists a(a, x \in A \rightarrow a \geqslant x \in C)$	A maximal element may not be unique. If "a" and "b" are both minimums or maximums in R, then $a \not\Vert b$ in R. Every finite partial hierarchy has at least one maximal and at least one minimal element. An infinite hierarchy need not have a maximal or minimal element [179].
Similarity	$\simeq$	$A \simeq B$: The hierarchy of "A" is the same as the hierarchy of "B."
Lower bound	$\forall x \exists m(x \in B \land m \in A \rightarrow m \leqslant x)$	If $B \subset A$, then "m" is a "lower bound" of "B."
Upper bound	$\forall x \exists m(x \in B \land m \in A \rightarrow m \geqslant x)$	If $B \subset A$, "m" is an "upper bound" of "B."

Hierarchies (Continued)		
Description	**Statement**	**Remark**
Greatest lower bound	$\forall m \forall x \exists M (x \in B \wedge m, M \in A \wedge m \leqslant x \rightarrow M \succ m)$	If $B \subset A$, then "M" is the "greatest lower bound" of "B" or "$inf(B)$."
Least upper bound	$\forall m \forall x \exists M (x \in B \wedge m, M \in A \wedge m \geqslant x \rightarrow M \prec m)$	If $B \subset A$, then "M" is the "least upper bound" of "B" or "$sup(B)$." There is at most one $inf(B)$ and one $sup(B)$ in hierarchies.

Chapter 1

God's Logic

1-1.6 And God said, "Let there be a firmament in the midst of the waters, and let it divide the waters from the waters."

— *The First Book of Moses*

An angel of God came to me in a dream. Her light was bright, and I trembled. She said, "Do not fear. I have something for you."

"What is it?" I asked.

"It is the logic of God and the creation of the universe. God commands you to give it to Man. You are to tell Man his story again, for he has forgotten it. And because he has forgotten his story, he is angry and unhappy. God is merciful and knows that Man is angry and unhappy and has suffered enough. So, go forth and teach Man his story again so he can be happy."

1.1 The Syntax of God's Logic

A set "S" will be called "Man's set." God gave Man a brain and "free will." Since God has "free will" and He created Man in His image, Man also has "free will." A set "$\bar{S}$" will be called "God's set." God gave Man a mind so he could discover who he is and know his story.

1.1.1 God's Language

If $s \in S$, then s signifies an idea in a man's brain of who he thinks he is. If $\bar{s} \in \bar{S}$, then $\bar{s}$ signifies an idea in a man's mind of how God made him. Thus,

$$\forall s \forall \bar{s}(s \in S \land \bar{s} \in \bar{S} \leftrightarrow \bar{s} \notin S \land s \notin \bar{S})$$

So, the thoughts in a man's brain are different from the thoughts in his mind. A man has "free will" and has his own thoughts; therefore, a man can be confused about who he is. To know who he is, a man must make a connection between who he thinks he is and how God intended him to be.

A relationship "r" represents a connection between a man's thoughts about himself, denoted by "S," and how God made him, denoted by "$\bar{S}$," written "$S \xrightarrow{r} \bar{S}$." So, S represents the body of a man, and $\bar{S}$ represents his mind. And "$S \xrightarrow{r} \bar{S}$" represents the man since he is composed of a body and a mind, which God gave him.

Example (see Fig. 1.1-1): Let $S = \{s_1, s_2, s_3, s_4\}$ and $\bar{S} = \{\bar{s}_1, \bar{s}_2, \bar{s}_3\}$, then r connects S with $\bar{S}$ by joining s_1 to $\bar{s}_1$ and s_2 to $\bar{s}_1$ and so on.

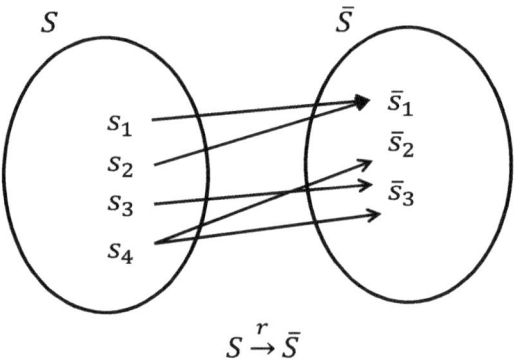

Figure 1.1-1

1.1.2 Statements of Humans about Themselves

The form "$r(s_i) = \bar{s}_j$" is a "statement" by a man about who he is. By fiat, "$r(s_i) = \bar{s}_i$" represents a "true" statement, and the man knows something real about himself. Otherwise, the man is deceived about who he is. For instance, in Fig. 1.1-1, the statement "$r(s_1) = \bar{s}_1$" is true, but "$r(s_2) = \bar{s}_1$" is false. A man who thinks false thoughts about himself is deceived.

1.1.3 Humans

The collection of statements "$S \xrightarrow{r} \bar{S}$" is what a man knows about himself. Similarly, the collection of statements by a woman is what she knows about herself. The collection of statements represents the "human." Some of the statements may be true, but others may be false.

1.1.3.1 To Thine Own Self be True

A man "$S \xrightarrow{r} \bar{S}$" knows his "true self" if

$$\forall s_i \forall \bar{s}_i (s_i \in S, \bar{s}_i \in \bar{S} \to r(s_i) = \bar{s}_i), \qquad i = 1,2,3,\ldots, \qquad \#S = \#\bar{S}$$

There is only one true "him" within all the ways a man can think about himself, which happens if:

1. r is one-to-one and onto.
2. All the statements about "him" are true.

If r is one-to-one and onto, then each thought a man has about who he is connects with one and only one thought in his mind. Moreover, "$\#S = \#\bar{S}$" means the two sets, Man's set "S" and God's set "$\bar{S}$" have the same number of thoughts.

1.1.3.2 Human Deception

If $S \xrightarrow{r} \bar{S}$ is a human and

$$\exists s_i \exists s_j (s_i, s_j \in S \to r(s_i) = r(s_j) = \bar{s}_k \in \bar{S}, \qquad i \neq j),$$

then such statements comprise a "confusion." A man confused about who he is suffers; his actions are wasted because he does not know who he is or where he fits. A man confused about who he is will be sorrowful and aggrieved.

Conversely, if $S \xrightarrow{r} \bar{S}$ is a human and

$$\exists \bar{s}_i \exists \bar{s}_j (\bar{s}_i, \bar{s}_j \in \bar{S} \to \bar{r}(\bar{s}_i) = \bar{r}(\bar{s}_j) = s_k \in S, \qquad i \neq j),$$

then such statements comprise a "confounding." A man confounded about who he is also suffers.

The less confounded and confused a man is about who he is, the more joy he feels because he will act in the way God intended. And the more confounded and confused a man is about who he is, the more sorrowful he will become. The man afflicted by sorrow will hate life because he will feel out of place.

1.1.3.3 Sense of Completeness and Incompleteness

The number of ways a man can think about who he is if S contains n thoughts and $\bar{S}$ contains m thoughts, is 2^{nm}, since there are two sets "$S, \bar{S}$." Such is the total number of ways thoughts in S can relate to thoughts in $\bar{S}$. A human senses an "incompleteness," however, if a thought in S does not relate to a thought in $\bar{S}$.

A man has a "complete" impression of himself if

$$\forall s_i \exists \bar{s}_j (s_i \in S \land \bar{s}_j \in \bar{S} \rightarrow r(s_i) = \bar{s}_j)$$

and

$$\forall \bar{s}_i \exists s_j (s_j \in S \land \bar{s}_i \in \bar{S} \rightarrow \bar{r}(\bar{s}_i) = s_j)$$

In other words, for every thought a man has of himself, the man will have at least one thought in his mind of who he is. He will feel a sense of "wholeness." A man can sense completeness but not know his "true" self (see Fig. *1.1-1*). A man who feels complete does not want and will not covet another's possessions. And he will wish others to be happy because he is complete and does not want.

A human "$S \xrightarrow{r} \bar{S}$" will sense "worthlessness" if

$$\forall s_i \exists \bar{s}_j (s_i \in S \land \bar{s}_j \in \bar{S} \rightarrow r(s_i) \neq \bar{s}_j)$$

Worthlessness is sadness a man feels about not knowing who he is. A man whose knowledge of himself is incomplete will covet others' things, feel depressed and lonely, and will seek unhappiness for others because he wants.

A man's knowledge about himself "$S \xrightarrow{r} \bar{S}$" is "fake" if

$$\exists s_i \forall \bar{s}_j (s_i \in S \land \bar{s}_j \in \bar{S} \rightarrow r(s_i) \neq \bar{s}_j)$$

A man who fakes knowledge about himself will have a higher or lower opinion of himself than he should. Such a man will suffer greed or hopelessness and self-righteousness or self-loathing. He will covet others' things because his knowledge of himself is incomplete, but he will never be satisfied.

1.2 Joy and Sadness

A man "$S \xrightarrow{r} \bar{S}$" will feel whole if his understanding of himself is complete. He will not covet. Wholeness, however, is a necessary but not a sufficient condition for the highest happiness.

A man thinks "consistently" about himself if each thought in S relates to one and only one thought in $\bar{S}$. Hence, if S contains n thoughts and $\bar{S}$ contains n thoughts, there are $n!$ complete and consistent ways a man can think about who he is.

1.2.1 Joy

A human "$S \xrightarrow{r} \bar{S}$" attains the highest "happiness" if

$$\forall s_i \forall \bar{s}_i (s_i \in S, \bar{s}_i \in \bar{S} \to r(s_i) = \bar{s}_i),$$

In other words, a man is happiest if he knows his "true" self.

1.2.2 Sadness

A human attains the highest "despondency" if

$$\forall s_i \forall \bar{s}_i (s_i \in S, \bar{s}_i \in \bar{S} \to r(s_i) \neq \bar{s}_i)$$

In other words, a man who is entirely ignorant of himself will be sad and will suffer mental illness.

1.2.3 Vacillation between Joy and Sadness

A human is sometimes happy and sometimes sad if

$$\exists s_i \exists \bar{s}_i (s_i \in S, \bar{s}_i \in \bar{S} \to r(s_i) = \bar{s}_i)$$

A man will suffer fluctuations in his thoughts if he does not know himself, but a man who gains a better understanding of his "true" self will rejoice and feel joy. A man who acquires a lesser knowledge of his "true" self will suffer and feel sadness. A happy man is productive, and a sad one wastes his life. A man should listen only to his voice about who he is and not the voices of others.

1.3 Human Relationships

If $A \xrightarrow{r} \bar{A}$ and $B \xrightarrow{f} \bar{B}$ are humans, then if

$$\langle x, y \rangle \in A \times B, \qquad x \in A, \qquad y \in B,$$

the ordered pair "$\langle x, y \rangle$" is called a "relation" between humans "$A \xrightarrow{r} \bar{A}$" and "$B \xrightarrow{f} \bar{B}$," where x is a thought from the brain of A and y is a thought from the brain of B.

If

$$\exists x \exists y (\langle x, y \rangle \in A \times B \to g[\langle x, y \rangle] = \langle \bar{x}, \bar{y} \rangle \in \bar{A} \times \bar{B}), \qquad \langle \bar{x}, \bar{y} \rangle \in \bar{A} \times \bar{B},$$
$$\bar{x} \in \bar{A}, \qquad \bar{y} \in \bar{B},$$

then the relationship is how God intended it and is "good." Otherwise, it is "bad."

1.3.1 Good and Bad Relationships

Given a system of relationships between two humans, if $C' \subset A \times B$ such that

$$\forall a \forall b (\langle a, b \rangle \in C' \subset A \times B \land \langle \bar{a}, \bar{b} \rangle \in \bar{C}' \to g[\langle a, b \rangle] = \langle \bar{a}, \bar{b} \rangle \in \bar{C}'),$$

then $C' \xrightarrow{g} \bar{C}'$ is the system of relationships God intended and is "good." If

$$\forall a \forall b (\langle a, b \rangle \in C' \subset A \times B \land \langle \bar{a}, \bar{b} \rangle \in \bar{C}' \to g[\langle a, b \rangle] \neq \langle \bar{a}, \bar{b} \rangle \in \bar{C}'),$$

then the system of relationships is entirely "bad." If

$$\exists a \exists b (\langle a, b \rangle \in C' \subset A \times B \land \langle \bar{a}, \bar{b} \rangle \in \bar{C}' \to g[\langle a, b \rangle] = \langle \bar{a}, \bar{b} \rangle \in \bar{C}'),$$

then the system of relationships will fluctuate between good and bad.

1.3.2 Relationships between Collections of Humans

Let

$$S = \exists x \exists y\{\langle x, y\rangle \in S_i \times S_j\}, \qquad \bar{S} = \exists \bar{x} \exists \bar{y}\{\langle \bar{x}, \bar{y}\rangle \in \bar{S}_i \times \bar{S}_j\},$$

where S is a system of relationships between a collection of humans "S_i" and a collection of humans "S_j", then $S \xrightarrow{r} \bar{S}$ relates two collections of humans to one another. A network of humans "$S \xrightarrow{r} \bar{S}$" is of God if

$$\forall x \forall y \left(\langle x, y\rangle \in S_i \times S_j \xrightarrow{r} \langle \bar{x}, \bar{y}\rangle \in \bar{S}_i \times \bar{S}_j \right)$$

In other words, the human relationships are the way God intended them to be.

1.3.3 Hierarchical Relationships

A "relation" is a hierarchy "$R = (A, A, C)$" if $a, b \in A$ and

$$\langle a, b\rangle \in A \times A \to a \leqslant b \in C$$

($a \leqslant b$: reads "a precedes b"). Hence, $a \leqslant b$ signifies a hierarchy on "A."

1.3.3.1 Similarity of Two Hierarchies

Suppose there are two hierarchies, "R" and "R'" respectfully. If $A \xrightarrow{f} B$, where every input into f has an output such that

$$\forall a \forall a' (a, a' \in A \wedge f(a), f(a') \in B \wedge a \leqslant a' \in C \to f(a) \leqslant f(a')),$$

then the hierarchies on A and B are "similar," designated "$A \simeq B$." If R is a hierarchy on A and $A \simeq B$, then R' and R are the same hierarchy. If $a \in A$ is a first (last, minimal, maximal) element and $A \simeq B$, then $f(a) \in B$ is a first (last, minimal, maximal) element in B.

1.4 Human Hierarchies

A man who understands the order and connection of his body's parts is happier than a man who does not.

Let $A \xrightarrow{f} \bar{A}$ be a human and let $R = (A, A, C)$ signify a hierarchy on A and $\bar{R} = (\bar{A}, \bar{A}, \bar{C})$ signify a hierarchy on $\bar{A}$. In other words, $\bar{R} = (\bar{A}, \bar{A}, \bar{C})$ is the hierarchy in which God arranged the body of A. If C is a system of hierarchical relations, i.e.,

$$\forall a \forall b (\langle a, b \rangle \in C \to a \leqslant b),$$

then $R \xrightarrow{g} \bar{R}$ is God's hierarchy if and only if $R \simeq \bar{R}$.

1.5 Human Paradoxes

God's logic does not permit a set of humans above all other sets of humans. A proof of this follows: Let $S = \{S_1, S_2, ...\}$ be the highest collection of humans. According to God's logic, S associates with a God set "$\bar{S} = \{\bar{S}_1, \bar{S}_2, ...\}$." But a set above all collections of humans requires that $\bar{S} \in S$, i.e.,

$$S = \{S_1, S_2, ..., \bar{S}_1, \bar{S}_2, ...\}$$

But each set in "S" associates with a God's set "$\bar{S}'$," i.e.,

$$\bar{S}' = \{\bar{S}_1, \bar{S}_2, ..., \bar{\bar{S}}_1, \bar{\bar{S}}_2, ...\}$$

But then the collection "S" would contain elements in "$\bar{S}'$," which God does not allow.

1.6 Concluding Remarks

God creates each man and woman differently, so the order and connection of their bodies' parts are a bit different. The more a man or a woman knows the order and relationship of their bodies' components, the more they know God, and the happier they will be. The more they learn about their uniqueness, the more they can offer Mankind.

No human should suppress another human. The hierarchy of society should be such that each human maximizes his or her potential. A suppressed human is a wasted life, and Mankind suffers.

A man or a woman should enter the relationships God intended for them and avoid bad relationships.

Each man and woman should learn to focus their mind on one thing for long periods and avoid distractions. The longer a man or a woman can focus on the disposition of his or her body, the more they will learn the order and connection of the parts of the body, the happier they will be.

A man or a woman who cannot focus and is ignorant of the true nature of his or her body will suffer "boredom." His or her fate will be unhappiness, uneasiness, impatience, discontent, and depression. He or she will seek instant gratification from sources outside of themselves but will not find it. Their boredom will grow, and they will seek higher levels of excitement from outside of themselves to relieve the boredom. A man or a woman in a perpetual state of boredom is dangerous because their thoughts will be of murder or suicide.

Learning the true nature of the disposition of the body takes effort. Do not look for instant solutions, but exercise self-discipline and patience. Concentration, self-discipline, patience, and effort are the seat of Man's powers. Through them, Man understands who he is and why he is.

1-1.26 And God said, "Let us make man in our image, after our likeness: and let them have dominion over the fish of the sea, and over the birds of the heavens, and over the cattle, and over all the earth, and over every creeping thing that creepeth upon the earth." 27 God created man in His own image, in the image of God He created him; male and female He created them.

The mind of human society is called the "government." There should be only one government—Man's government, ruled by God. If there is only one government, there will be no more war because the government has no one to fight against. When there are many governments, they will war against each other to become the only government. However, when there is only one government, it should no longer spend its resources on war but for the betterment of Mankind.

The government's hierarchy should function as the parts of the human body because that is what God intended. Each man and each woman who knows himself or herself should find their place in the hierarchy, whether they are part of the foot, the legs, the arms, or the head. Each has its purpose. The hierarchy of government should be such that each man and woman can find their place in the body of Mankind because that is what God intended.

The hierarchy of government that serves only the mind of Man is "capitalism." It is an incomplete system, where Man fabricates the importance of things and places himself above God. Driven by greed and self-righteousness, men think too highly of themselves. This hierarchy of government will end in destruction. It is not the right government for Man.

The hierarchy of government that serves only the body of Man is "socialism." It is an incomplete system, where Man places himself below God. Driven by melancholy, men think too little of themselves. This hierarchy of government will end in destruction. It is not the right government for Man.

"Communism" is the hierarchy of Socialism but forced upon Man by men who think they are superior to other men. It will drive men to anger, distrust, and sadness. It is malignant Socialism. It is not the right government for Man.

The government hierarchy that makes no distinction between the mind and the body is "Fascism." It is a complete system where Man replaces God. Driven by death, men become machines. This hierarchy of government will end in destruction. It is not the right government for Man.

The hierarchy of government for Man should divide Man into "thinkers" and "doers," for this is how God divided Man, mind, and body. The thinkers produce the policies for Man, and the doers execute the strategies. But most men and women will have some of both traits and must decide where they best fit.

God divided the darkness from the light. So, God made man and woman. But man and woman are both mind and body, both dark and light. The man's body is primarily dark because it cannot produce life. Only a woman's body can do that. So, in the government hierarchy, the doers should be mostly women, although this will not always be the case.

The man's body is primarily dark, but his mind is mostly light. The woman's mind is predominantly dark, so she holds the mind's seed, but the man creates with his mind. So, in the government hierarchy, the thinkers should be mostly men, although this will not always be the case.

A woman's mind should keep busy; otherwise, she could contemplate death. The man should have the best intentions for the woman so she can be happy instead of depressed.

In all human endeavors, there is theory and practice, the mind, and the body. Theory and practice should be the same. So, each man and woman should seek their place, whether in the mind or the body of Mankind.

The seat of government for Man should be Jerusalem. For man came from there, and it is there he should return. It is his home.

Chapter 2

God's Numbers

"God created everything by number, weight, and measure."

— *Sir Isaac Newton*

God's numbers follow from God's logic. The set "S" will contain "Man's numbers." The set "$\bar{S}$" contains "God's numbers." God made numbers so Man could discover them. So, God placed numbers in the mind of Man so he could find them. The relationship "$S \xrightarrow{r} \bar{S}$" is called a "number system," which shows the connection between Man's numbers and God's numbers.

2.1 Natural Numbers

The set "$\mathbb{N} = \{1,2,3,...\}$" is Man's "natural numbers." The set "$\bar{\mathbb{N}}$" is God's "natural numbers." The number system "$\mathbb{N} \xrightarrow{r} \bar{\mathbb{N}}$" shows how Man's natural numbers relate to God's natural numbers.

Suppose $\bar{\mathbb{N}} = \{-1, -2, -3, ...\}$, where

$$\forall n (n \in \mathbb{N} \to r(n) = -n = \bar{n} \in \bar{\mathbb{N}}), \qquad n = 1, 2, 3, ...$$

In the set of statements "$r(n) = -n$," each n is associated with a unique $-n$. Hence, $\mathbb{N} \xrightarrow{r} \bar{\mathbb{N}}$ is consistent and complete. All the statements "$r(n) = \bar{n}$" are, by fiat, true. Therefore, $\mathbb{N} \xrightarrow{r} \bar{\mathbb{N}}$ is a number system. God's natural numbers are such that multiplying two negative numbers results in a negative number, i.e.,

$$-m \cdot (-n) = -mn$$

Man did not know this, but he knows it now.

Moreover, $\mathbb{N}$ has a hierarchy, i.e.,

$$\forall n (n \in \mathbb{N} \to n < n + 1)$$

and ℕ has a first element "1" so that $r(1) = -1$. If "1" is the first element in ℕ, then "−1" is the first element in $\bar{ℕ}$ and

$$\forall n(n \in ℕ, n < n + 1 \rightarrow r(n) < r(n+1) \rightarrow -n < -n - 1 \in \bar{ℕ})$$

Hence, $\bar{ℕ} \simeq ℕ$. Therefore, $(ℕ, \mathbb{N}, C) \xrightarrow{r} (\bar{ℕ}, \bar{\mathbb{N}}, \bar{C})$ constitutes God's hierarchy.

Man has discovered the system of natural numbers, i.e.,

$$ℕ \xrightarrow{r} \bar{ℕ}$$

2.2 Integers

The set "$\mathbb{Z} = \{..., -3, -2, -1, 0, 1, 2, 3, ...\}$" is "Man's integers." The set "$\bar{\mathbb{Z}}$" is "God's integers." The number system "$\mathbb{Z} \xrightarrow{r} \bar{\mathbb{Z}}$" shows how Man's integers relate to God's integers.

Define

$$\sqrt{\mathbb{Z}} = \{\pm\sqrt{n}\} \rightarrow \mathbb{Z} = \{\pm\sqrt{n} \cdot \sqrt{n} = \pm n\}, \quad n \in ℕ, \quad n = 1, 2, 3, ...,$$
$$0 \in \mathbb{Z},$$

which are Man's integers. They come from Man's natural numbers. Let

$$\sqrt{\bar{\mathbb{Z}}} = \{\pm\sqrt{-n}\} \rightarrow \bar{\mathbb{Z}}$$
$$= \{\pm\sqrt{-n} \cdot \sqrt{-n} = \pm\sqrt{(-n) \cdot (-n)} = \pm\sqrt{-(n \cdot n)}$$
$$= \pm\sqrt{-1} \cdot \sqrt{n \cdot n} = \pm in\}, \quad -n \in \bar{ℕ}, \quad i = \sqrt{-1},$$
$$i0 = \bar{0} \in \bar{\mathbb{Z}},$$

which are God's integers, where "in" is called an "imaginary number." They come from God's natural numbers, which creates two number systems "$\mathbb{Z} \xrightarrow{f} \bar{\mathbb{Z}}$." In the first case, $f(n) = in$ for all n, where $\pm n \in \mathbb{Z} \sim 0$ and $\pm in \in \bar{\mathbb{Z}} \sim \bar{0}$. A set of ordered pairs can represent each system. The pair "$\langle n, n \rangle$" lies along the upward sloping line in Fig. 2.2-1. In the second case, $f(-n) = in$ for all n. Each pair "$\langle -n, n \rangle$" lies along the downward sloping line in Fig. 2.2-1. The rules are now

$$(-m) \cdot (-n) = mn, \quad (m) \cdot (-n) = -mn, \quad m, n \in \mathbb{Z}$$

And

$$(-im) \cdot (-in) = imn, \quad (im) \cdot (-in) = -imn, \quad im, in \in \overline{\mathbb{Z}}, \quad i \cdot i = i$$

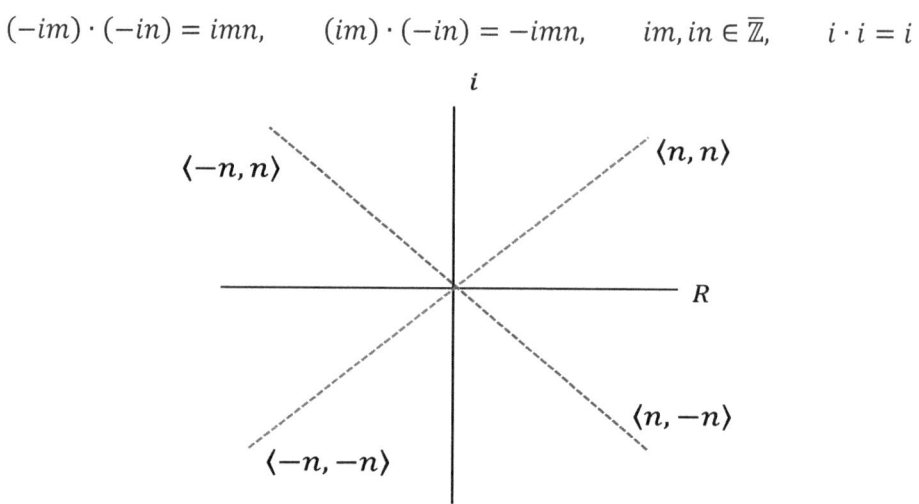

Figure 2.2-1

In both cases, each n maps to a unique in for all n. Both systems are complete and consistent. In other words, in each system, considered separately,

$$\forall n(n \in \mathbb{Z} \sim 0 \rightarrow f(n) = \bar{n} \in \overline{\mathbb{Z}} \sim \bar{0})$$

Hence, $\mathbb{Z} \sim 0 \xrightarrow{f} \overline{\mathbb{Z}} \sim \bar{0}$ is a number system in each case. The set "$\mathbb{Z} \sim 0$" is a total hierarchy, but without a maximum or minimum number, by the ordering in $\mathbb{N}$:

$$\forall n(n < n+1 \in \mathbb{Z} \sim 0 \rightarrow f(n) < f(n+1) = \bar{n} < \bar{n} + \bar{1} \in \overline{\mathbb{Z}} \sim \bar{0})$$

Since $\mathbb{Z} \sim 0 \simeq \overline{\mathbb{Z}} \sim \bar{0}$ in both systems, the systems are God's hierarchies.

Finally, define a number "$0 \in \mathbb{Z}$" such that $f(0) = i0 = \bar{0} \in \overline{\mathbb{Z}}$. The pair "⟨0,0⟩" is a member of both the upward and downward sloping systems (see Fig. 2.2-1) and is the only pair belonging to both systems. There are two systems of integers:

$$\mathbb{Z} \xrightarrow{r} \overline{\mathbb{Z}},$$

Man did not know this, but he knows it now.

2.3 Rational Numbers

The set "$\mathbb{Q} = \{n/m, 0\}$," $n, m \in \mathbb{Z}{\sim}0$ is Man's "rational numbers." They come from Man's integers. God's "rational numbers" are these:

$$\overline{\mathbb{Q}} = \{i\frac{n}{m}, i0\}, \quad i = \sqrt{-1}, \quad i \cdot i = i$$

Like the integers, there are two number systems "$\mathbb{Q} \xrightarrow{f} \overline{\mathbb{Q}}$." In the first case,

$$\forall m \forall n \left(m, n \in \mathbb{Z}{\sim}0 \rightarrow f\left(\frac{n}{m}\right) = i\frac{n}{m} \in \overline{\mathbb{Q}}\right), \quad f(0) = \overline{0},$$

where each pair "$\langle n/m, n/m \rangle$" lies along the upward sloping line in Fig. 2.2-1. In the second case,

$$f\left(-\frac{n}{m}\right) = i\frac{n}{m}, \quad f(0) = \overline{0},$$

where each pair "$\langle -n/m, n/m \rangle$" lies along the downward sloping line. Like $\mathbb{Z}$, multiplying two imaginary rational numbers results in an imaginary rational number, i.e., $i \cdot i = i$.

The systems of rational numbers have a hierarchy:

$$\forall \bar{x} \forall \bar{y} (\bar{x}, \bar{y} \in \overline{\mathbb{Q}} \wedge \bar{x} < \bar{y} \leftrightarrow \bar{x} - \bar{y} < \overline{0}),$$

$$\forall \bar{x} \forall \bar{y} (\bar{x}, \bar{y} \in \overline{\mathbb{Q}} \wedge \bar{x} > \bar{y} \leftrightarrow \bar{x} - \bar{y} > \overline{0}),$$

$$\forall \bar{x} \forall \bar{y} (\bar{x}, \bar{y} \in \overline{\mathbb{Q}} \wedge \bar{x} = \bar{y} \leftrightarrow \bar{x} - \bar{y} = \overline{0})$$

The set "$\mathbb{Q}$" has the same hierarchy as $\overline{\mathbb{Q}}$. Hence, $\mathbb{Q} \simeq \overline{\mathbb{Q}}$. There are two systems of rational numbers:

$$\mathbb{Q} \xrightarrow{r} \overline{\mathbb{Q}}$$

Man did not know this, but he knows it now.

2.4 Real Numbers

The real numbers "$\mathbb{R}$" include both rational and irrational numbers. Irrational numbers are those that are not integers nor fractions. Although both sets of numbers, the rational and irrational numbers, are infinite, there are more irrational than rational numbers. Irrational numbers are broken into two groups, "algebraic" and "transcendental." Man knows most rational and algebraic irrational numbers. But Man only knows a few transcendental numbers.

The systems "$\mathbb{R} \xrightarrow{f} \overline{\mathbb{R}}$" develop similarly to the rational numbers. God's "$\overline{\mathbb{R}}$" relates to Man's "$\mathbb{R}$" in roughly the same way $\overline{\mathbb{Q}}$ relates to $\mathbb{Q}$:

$$\mathbb{R} \xrightarrow{f} \overline{\mathbb{R}},$$

the real number systems.

2.5 Complex Numbers

Complex numbers "$\mathbb{C}$" are two-dimensional (see Fig. 2.5-1).

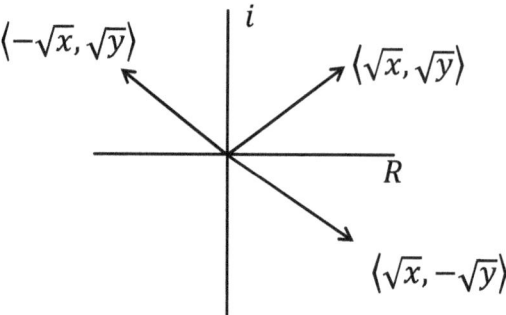

Figure 2.5-1

Man's "complex numbers" consist of the ordered pairs "$\langle x, y \rangle$", $\langle x, y \rangle \in \mathbb{C}$, $x, y \in \mathbb{R}$. The second and fourth quadrants in Fig. 2.5-1 are reserved for "God's numbers."

To see this, let

$$\sqrt{\mathbb{C}} = \{\langle\pm\sqrt{x},\pm\sqrt{y}\rangle\}, \qquad \sqrt{x},\sqrt{y} \in \{\sqrt{\mathbb{R}_+},0\}, \qquad x,y \geq 0 \in \mathbb{R},$$

which implies that

$$\mathbb{C} = \{\langle\pm\sqrt{x}\sqrt{x},\pm\sqrt{y}\sqrt{y}\rangle\} = \{\langle\pm x,\pm y\rangle\}, \qquad x,y \geq 0 \in \mathbb{R}$$

Let

$$\sqrt{\mathbb{R}} = \{\pm\sqrt{x}\}, \qquad \overline{\sqrt{\mathbb{R}}} = \{\pm i\sqrt{y}\}, \qquad x,y \geq 0 \in \mathbb{R}, \qquad i = \sqrt{-1}$$

Define

$$\{\langle\sqrt{\mathbb{R}},\overline{\sqrt{\mathbb{R}}}\rangle\} = \{\langle\pm\sqrt{x},\pm\sqrt{y}\rangle\} = \sqrt{\mathbb{C}} \to \{\langle\mathbb{R},\overline{\mathbb{R}}\rangle\} = \{\langle\pm x,\pm y\rangle\} = \mathbb{C}$$

The set "$\mathbb{C}$" is Man's "complex numbers." The rule is now

$$i^2 = -1, \qquad i = \sqrt{-1}$$

The rule for adding two of Man's complex numbers "$\langle a,b\rangle$" and "$\langle c,d\rangle$" is

$$\langle a,b\rangle + \langle c,d\rangle = \langle a+c, b+d\rangle, \qquad a,b,c,d \in \mathbb{R}$$

The rule for multiplying two of Man's complex numbers "$\langle a,b\rangle$" and "$\langle c,d\rangle$" is

$$\langle a,b\rangle \cdot \langle c,d\rangle = \langle ac-bd, ad+bc\rangle$$

2.5.1 Number Entanglement

The pair "$\langle\sqrt{x},-\sqrt{y}\rangle \in \sqrt{\mathbb{C}}$" is the "conjugate" of $\langle\sqrt{x},\sqrt{y}\rangle$. Hence,

$$\langle\sqrt{x},-\sqrt{y}\rangle \to \langle\sqrt{x}\sqrt{x},-\sqrt{y}\sqrt{y}\rangle = \langle x,-y\rangle \in \mathbb{C}$$

Define the "anti-conjugate" as

$$\langle-\sqrt{x},\sqrt{y}\rangle \in \sqrt{\mathbb{C}} \to \langle-\sqrt{x}\sqrt{x},\sqrt{y}\sqrt{y}\rangle = \langle-x,y\rangle \in \mathbb{C},$$

so that

$$\langle x, -y \rangle + \langle -x, y \rangle = \langle 0, 0 \rangle$$

Now

$$\langle \sqrt{x}, \sqrt{y} \rangle \cdot \langle \sqrt{x}, -\sqrt{y} \rangle = \langle x+y, 0 \rangle = r \to \sqrt{r} = \pm\sqrt{x+y} \in \sqrt{\mathbb{R}}$$
$$\to \langle x, y \rangle \cdot \langle x, -y \rangle = x^2 + y^2 = r_R^2 \to r_R = \pm\sqrt{x^2+y^2} \in \mathbb{R}$$

On the other hand,

$$\langle \sqrt{x}, \sqrt{y} \rangle \cdot \langle -\sqrt{x}, \sqrt{y} \rangle = \langle -x-y, 0 \rangle = -(x+y) = -r \to i\sqrt{r} = \pm i\sqrt{x+y}$$
$$\in \overline{\sqrt{\mathbb{R}}} \to \langle x, y \rangle \cdot \langle -x, y \rangle = -x^2 - y^2 = -r^2 = -(x^2+y^2)$$
$$\to ir_R = \pm i\sqrt{x^2+y^2} \in \overline{\mathbb{R}}$$

Let $i\sqrt{y}$ be Man's part of the imaginary part of a complex number. And let $-i\sqrt{y}$ be God's part (Fig. 2.5-1). In other words,

$$\forall \sqrt{y} \exists f\left(i\sqrt{y} \in I_m(\sqrt{\mathbb{C}})\right) \to f(i\sqrt{y}) = -i\sqrt{y}, \qquad y \geq 0 \in \mathbb{R},$$

where "$I_m(\sqrt{\mathbb{C}})$" is the imaginary part of a complex number.

To make the real part of God's complex number, use the anti-conjugate:

$$\langle -\sqrt{x}, \sqrt{y} \rangle \to \forall \sqrt{x} \exists f\left(\sqrt{x} \in R_e(\sqrt{\mathbb{C}})\right) \to f(\sqrt{x}) = -\sqrt{x}, \qquad x \geq 0 \in \mathbb{R},$$

where "$R_e(\sqrt{\mathbb{C}})$" is the real part of a complex number.

Let $\sqrt{r} = \sqrt{x+y}$ associate with $i\sqrt{y}$, but not $\sqrt{x}$ and $i\sqrt{r} = i\sqrt{x+y}$ associate with $\sqrt{x}$, but not $i\sqrt{y}$. Therefore, for the real part of a complex number system,

$$R_e\langle \pm\sqrt{x}, \pm\sqrt{y} \rangle = \pm\sqrt{x} \overset{g}{\to} \pm i\sqrt{r} = \pm i\sqrt{x+y},$$

where g sends $\pm\sqrt{x}$ into $\pm i\sqrt{x+y}$.

And, for the imaginary part,

$$I_m\langle \pm\sqrt{x}, \pm\sqrt{y} \rangle = \pm i\sqrt{y} \overset{g}{\to} \pm\sqrt{r} = \pm\sqrt{x+y},$$

where g sends $\pm i\sqrt{y}$ into $\pm\sqrt{x+y}$.

Let $\sqrt{x} \to i\sqrt{y}$, $i\sqrt{y} \to \sqrt{x}$, then

$$\pm i\sqrt{y} \overset{g}{\to} \pm i\sqrt{r} = \pm i\sqrt{x+y}, \qquad \pm\sqrt{x} \overset{g}{\to} \pm\sqrt{r} = \pm\sqrt{x+y}$$

Hence, let

$$\langle \pm x, \pm y \rangle \overset{g}{\to} \langle \pm\sqrt{x}\sqrt{x+y}, \pm\sqrt{y}\sqrt{x+y} \rangle, \qquad x, y \geq 0 \in \mathbb{R},$$

where g sends $\pm x$ into $\pm\sqrt{x}\sqrt{x+y}$ and $\pm iy$ into $\pm i\sqrt{y}\sqrt{x+y}$.

2.5.1.1 Complex Number Systems

The set "$\mathbb{C} = \{\langle \pm x, \pm y \rangle\}$" is Man's "complex numbers." The set of God's complex numbers is

$$\bar{\mathbb{C}} = \{\langle \pm\sqrt{x}\sqrt{x+y}, \pm\sqrt{y}\sqrt{x+y} \rangle\},$$

i.e.,

$$\langle \pm x, \pm y \rangle \overset{f}{\to} \langle \pm\bar{x}, \pm\bar{y} \rangle, \qquad \bar{x} = \sqrt{x}\sqrt{x+y}, \qquad \bar{y} = \sqrt{y}\sqrt{x+y},$$
$$x, y \geq 0 \in \mathbb{R},$$

where $\langle x, y \rangle \in \mathbb{C}$ and $\langle \sqrt{x}\sqrt{x+y}, \sqrt{y}\sqrt{x+y} \rangle \in \bar{\mathbb{C}}$. The system "$\mathbb{C} \overset{f}{\to} \bar{\mathbb{C}}$" is a complex number system (see Fig. 2.5.1-2).

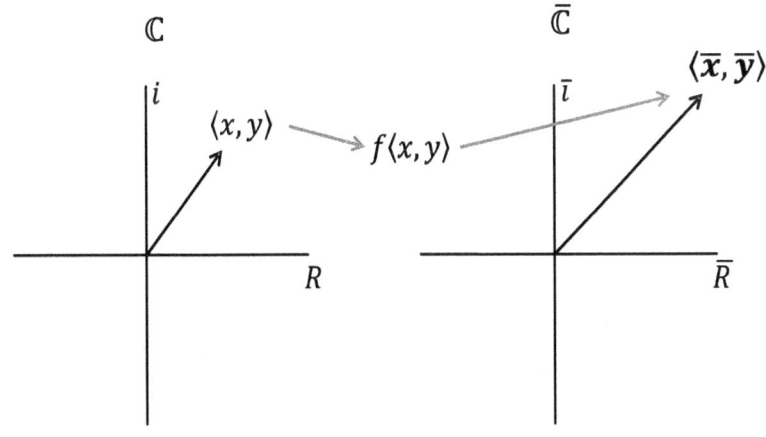

Figure 2.5.1-2

Moreover, let

$$\langle \sqrt{x}\sqrt{x+y}, \sqrt{y}\sqrt{x+y}\rangle \xrightarrow{g} \langle \sqrt{y}\sqrt{x+y}, -\sqrt{x}\sqrt{x+y}\rangle,$$

where the right-hand side comes from multiplying the left-hand side by $-i$ and g sends $\sqrt{x}\sqrt{x+y}$ into $\sqrt{y}\sqrt{x+y}$ and $\sqrt{y}\sqrt{x+y}$ into $-\sqrt{x}\sqrt{x+y}$. In other words,

$$\langle x, y\rangle \xrightarrow{g \circ f} \langle \sqrt{y}\sqrt{x+y}, -\sqrt{x}\sqrt{x+y}\rangle$$

is also a complex number system.

If

$$\langle \sqrt{x}\sqrt{x+y}, \sqrt{y}\sqrt{x+y}\rangle \in \bar{\mathbb{C}},$$

then the real and imaginary parts of God's complex numbers are not independent. If $x \neq 0$, an exact value for x only happens if

$$\sqrt{x}\sqrt{x+y} = \sqrt{x}\sqrt{x} = x, \qquad y = 0$$

Conversely, an exact value for $y \neq 0$ happens only if $x = 0$. Such a dependency is called "number entanglement" [282]. Note that if x and y have dimension "d," then

$$\sqrt{x\,[d]}\sqrt{x\,[d] + y\,[d]} = \sqrt{x}\,[\sqrt{d}]\sqrt{x+y}\,[\sqrt{d}] = \sqrt{x}\sqrt{x+y}\,[\sqrt{d}\sqrt{d}]$$
$$= \sqrt{x}\sqrt{x+y}\,[d]$$

In other words, $\sqrt{x}\sqrt{x+y}$ comes in whole units, as does $\sqrt{y}\sqrt{x+y}$.[1]

2.6 Concluding Remarks

These are the relationships between God's and Man's numbers. God used complex numbers to create the universe. Man has not known the relationship between his complex numbers and God's complex numbers, but he knows now.

---

[1] See Appendix "A1" for operations on complex numbers.

Chapter 3

How God Made the Universe

"1-1.1 In the beginning God created the heavens and the earth."

— The First Book of Moses

In the beginning, God divided the universe into the heavens and the earth. 1-1.6 And God said, "Let there be a firmament in the midst of the waters, and let it divide the waters from the waters." 7 And God made the firmament, and divided the waters which were under the firmament from the waters which were above the firmament: and it was so.

Mathematically, the function "f" will represent the "heavens:"

$$f(t) = \left(-\frac{1}{G}\right)^{1-t} G^t, \quad G \in \mathbb{R}$$

The heavens are the "large physics." The function "g" will represent "matter" (earth):

$$g(t) = \left(-\frac{1}{G}\right)^{t-1} (G)^{-t},$$

the "small physics." Now $fg = 1$, so there is only one universe—the one created by God. God called the function "f" the heavens. And he called the function "g" Lucifer because it was beautiful. Lucifer was the most delightful thing that God had made. God separated the waters from the waters.

1-1.2 And the earth was waste and void, and darkness was upon the face of the deep: and the Spirit of God moved upon the face of the waters.

The spirit is "time," represented by "t" in the functions above. God started the clock, which is how the universe began.

1-1.3 Then God said, "Let there be light."; and there was light.

For each value of t, f produces a complex number. It is the same with g, which is how God separated the light from the darkness. And God saw that it was good.

3.1 God's Functions

If $G = -1$, f produces a complex unit circle (see Fig. 3.1-1):

$$f(t) = \left(-\frac{1}{G}\right)^{1-t} G^t, \qquad f(t) = (-1)^t = \cos(\pi t) + i\sin(\pi t) = e^{i\pi t}$$

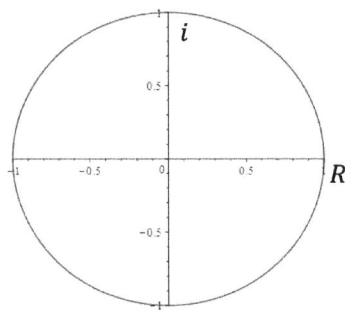

Figure 3.1-1

If $G = -\sqrt{10}\, e^{1/2}/10$, then f produces a complex spiral (see Fig. 3.1-2):

$$f(t) = 10^{1/2-t} e^{-1/2+t}[\cos(\pi t) + i\sin(\pi t)] = A(t)e^{i\pi t},$$
$$A(t) = 10^{1/2-t} e^{-1/2+t}$$

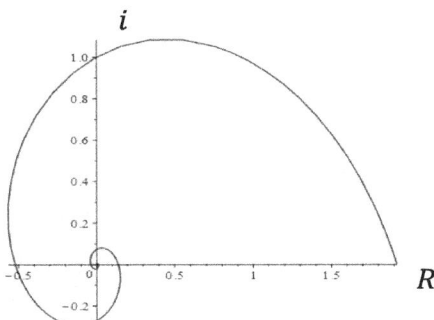

Figure 3.1-2

If $G = -1$, then g also makes a complex unit circle (see Fig. 3.1-3):

$$g(t) = \cos(\pi t) - i\sin(\pi t) = e^{-i\pi t}$$

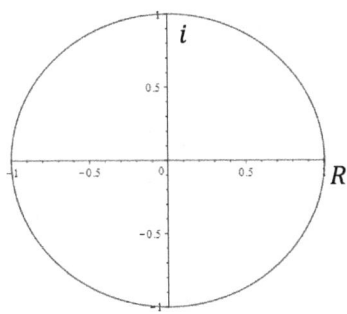

Figure 3.1-3

And if $G = -\sqrt{10}\, e^{1/2}/10$, then g makes a complex spiral on an enormous scale (see Fig. 3.1-4):

$$g(t) = 10^{(-1/2+t)} e^{(1/2-t)}[\cos(\pi t) - i\sin(\pi t)] = A(t)e^{-i\pi t},$$
$$A(t) = 10^{-1/2+t} e^{1/2-t}$$

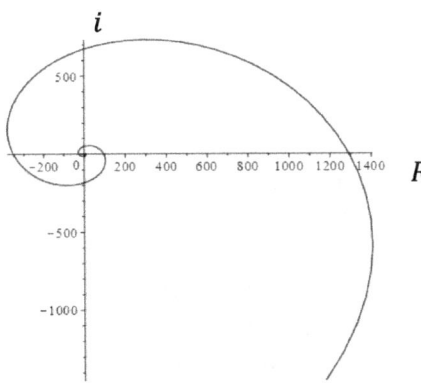

Figure 3.1-4

3.2 God

God is the unmoved mover. Two numbers represent Him, one that moves and one that does not move. The number that moves is:

$$e = \lim_{n \to \infty} \left(1 + \frac{1}{n}\right)^n, \quad n \in \mathbb{N}$$

The number "e" is called a "moving number" because it is the limit of an infinite number of numbers. God moves through the natural numbers. For each n, e becomes a different number. If $n = 1$, then $e = 2$. If $n = 2$, then $e = 2.25$. If $n = 3$, $e = 2.48832$, and so on. Mathematicians call "e" a "continuous" or "transcendental" number.

The number "10" is a rational number because it does not move. "G" is the number of God:

$$10^{15} G = e \to G = \frac{e}{10^{15}} \to \sqrt{G} = \pm\sqrt{\frac{e}{10^{15}}} \approx \pm 5.213714442 \times 10^{-8}$$

The number "10" is called a "discrete" number. So, God is both continuous and discrete.

Numerically, "$|\sqrt{G}|$" is approximately equal to the constant of proportionality associated with the Newtonian theory of gravity in the "CGS" [$centimeters - grams - seconds$] system of units [283]. Man knows God through Isaac Newton's theory of gravity. Newton's law of gravitation:

$$F = -|\sqrt{G}|\frac{m_1 m_2}{r^2}, \quad \sqrt{G} \approx \pm 5.213714442 \times 10^{-8},$$

i.e., two "masses" (m_1, m_2) attract by a "force" (F), proportional to the square of the "distance" (r) between them.

But God is in everything and is everywhere. The modern word for God is "consciousness." God is now and forever. God is eternal. He always was and always will be. He is ever-changing but never changes. He is nowhere but everywhere.

God wants Man to know Him again. Man must return to Eden. It is time for Man to return to God.

3.2.1 The Four Functions of God

God's number put into God's functions produces four functions:

$$\sqrt{f(t)} = \left(-\frac{1}{\sqrt{G}}\right)^{1-t} \left(\sqrt{G}\right)^{t}, \quad \sqrt{f_-(t)} = \left(\frac{1}{\sqrt{G}}\right)^{1-t} \left(-\sqrt{G}\right)^{t},$$

$$\sqrt{g(t)} = \left(-\frac{1}{\sqrt{G}}\right)^{t-1} \left(\sqrt{G}\right)^{-t}, \quad \sqrt{g_-(t)} = \left(\frac{1}{\sqrt{G}}\right)^{t-1} \left(-\sqrt{G}\right)^{-t}$$

Therefore, all of God's creations are "mind" and "body" (see Table 3.2.1-1)[2]. And God divides things into "male" and "female" or "positive" and "negative," each of its kind.

1-1.27 So, God created humankind in His image, in the image of God He created them; male and female He created them.

Therefore, a man has a mind and a body, as does a woman.

	Male	Female
Mind	$\sqrt{f(t)} = \left(-\frac{1}{\sqrt{G}}\right)^{1-t} \left(\sqrt{G}\right)^{t}$	$\sqrt{f_-(t)} = \left(\frac{1}{\sqrt{G}}\right)^{1-t} \left(-\sqrt{G}\right)^{t}$
Body	$\sqrt{g(t)} = \left(-\frac{1}{\sqrt{G}}\right)^{t-1} \left(\sqrt{G}\right)^{-t}$	$\sqrt{g_-(t)} = \left(\frac{1}{\sqrt{G}}\right)^{t-1} \left(-\sqrt{G}\right)^{-t}$

Table 3.2.1-1

3.2.2 The Birth of Man

1-2.7 And God formed man from the dust of the ground, and breathed into his nostrils the breath of life; and man became a living soul. 8 And God planted a garden eastward, in Eden; and there He put the man whom He had formed ... 16 And God commanded the man, saying, "Of every tree of

---

[2] Note that $\left(\sqrt{f(t)}\right)^2 \neq f(t)$.

the garden thou mayest freely eat: 17 but of the tree of the knowledge of good and evil, thou shalt not eat of it: for in the day that thou eat of it, thou shalt surely die."

1-2.18 And God said, "It is not good that the man should be alone; I will make him a helper ..." 21 And God caused a deep sleep to fall upon the man, and he slept, and He took one of his ribs and closed up the flesh. 22 And the rib, which God had taken from the man, made a woman and brought her unto the man. 23 And the man said, "This is bone of my bones, and flesh of my flesh: she shall be called Woman because she was taken out of the man."

Man's body consists of earth, his mind made from the heavens. Now when God made the heavens, He used f; He counted all time and created the speed that all bodies could not exceed:

$$\int \sqrt{f}\, d\omega = -5.669067105 \times 10^5 + i53102.41304$$

And God created both the dark-speed and the light-speed, for God separated the "day" from the "night." The square root of the speed of the dark "$\sqrt{x}$" is:

$$\sqrt{x} = -5.669067105 \times 10^5$$

and the square root of the speed of the light "$\sqrt{y}$" is:

$$\sqrt{y} = 53102.41304$$

Man only knows the speed of the light, but he should learn the speed of the dark:

$$x \text{ (Speed of the dark)} = 3.213832184 \times 10^{11}$$

$$y \text{ (Speed of the light)} = 2.819866266 \times 10^9$$

1-3.1 Now the serpent was more subtle than any beast of the field which God had made. And He said unto the woman, "Yea, hath God said, "Ye shall not eat of any tree of the garden?" 2 And the woman said unto the serpent, "Of the fruit of the trees of the garden we may eat: 3 but of the fruit of the tree which is in the midst of the garden, God hath said, 'Ye shall not eat of it, neither shall ye touch it, lest ye die.'" 4 And the serpent said unto the woman, "Ye shall not surely die: 5 for God doth know that in the day ye eat of it, then your eyes shall be opened, and ye shall be as God, knowing good and evil." 6 And when the woman saw that the tree was good for food, and that it was a delight to the eyes, and that the tree was to be desired to make one wise, she took of the fruit, and did eat; and she gave also unto her husband who was with her, and he did eat. 7 And the eyes of them both were opened, and they knew that they were naked; and they sewed fig leaves together, and made themselves aprons. 8 And they heard the voice of God walking in the garden in the cool of the day: and the man and his wife hid themselves from the presence of God amongst the trees of the garden. 9 And God called unto the man, and said unto him, "Where art thou?" 10 And he said, "I heard thy voice in the garden, and I was afraid, because I was naked; and I hid myself." 11 And He said, "Who told thee that thou wast naked? Hast thou eaten of the tree, whereof I commanded thee that thou shouldest not eat?" 12 And the man said, "The woman whom thou gavest to be with me, she gave me of the fruit, and I did eat."

1-3.13 And God said unto the woman, "What is this thou hast done?" And the woman said, "The serpent beguiled me, and I did eat." 14 And God said unto the serpent, "Because thou hast done this, cursed art thou above all cattle, and above every beast of the field; upon thy belly shalt thou go, and dust shalt thou eat all the days of thy life: 15 and I will put enmity between thee and the woman, and between thy seed and her seed: he shall bruise thy head, and thou shalt bruise his heel."

God had made Lucifer endless and beautiful. When God added up all the time for Lucifer:

$$\int \sqrt{g}\, dt = \int \left(-\frac{1}{\sqrt{G}}\right)^{t-1} \left(\sqrt{G}\right)^{-t} dt = \infty$$

Lucifer would grow endlessly and never know death. But because of his beauty, Lucifer became vain and tried to exalt himself above God. There was a war in Heaven, and God cast Lucifer and His minions to the earth so that he would grow humble and not vain anymore. Lucifer would seduce the woman. So, God made him small and renamed him "Satan," the

"Prince of Darkness." So, God took time away from Satan and made him finite. And He put him on the ground, and he became like dust:

$$3.111669924 \times 10^{-12}$$

is the devil's number, and God made Satan small. And Man's body, created from dust, will die because Satan is embodied in Man.

1-3.16 Unto the woman He said, "I will greatly multiply thy pain and thy conception; in pain thou shalt bring forth children; and thy desire shall be to thy husband, and he shall rule over thee." 17 And unto Adam He said, "Because thou hast hearkened unto the voice of thy wife, and hast eaten of the tree, of which I commanded thee, saying, 'Thou shalt not eat of it' cursed is the ground for thy sake; in toil shalt thou eat of it all the days of thy life; 18 thorns and thistles shall it bring forth to thee; and thou shalt eat the herb of the field; 19 in the sweat of thy face shalt thou eat bread, till thou return unto the ground; for out of it wast thou taken: for dust thou art, and unto dust shalt thou return."

1-3.22 And God said, "Behold, the man has become like one of us, to know good and evil; and now, lest he put forth his hand, and take also of the tree of life, and eat, and live forever." 23 Therefore God sent him forth from the garden of Eden, to till the ground from whence he was taken. 24 So He drove out the man; and He placed at the east of the garden of Eden the Cherubim, and the flame of a sword which turned every way, to guard the way to the tree of life.

So, God took the dark and the light "$\langle D, L \rangle$" and entangled them. So, when Man would look for the things in Eden, the dark and the light, he would not see them anymore but would see instead

$$\langle \sqrt{D}\sqrt{D+L}, \sqrt{L}\sqrt{D+L} \rangle$$

God cut Man off from seeing things as they are, so Man is confused and cannot see Eden or God no matter where he looks. Man yearns to see God but will not find Him.

3.2.3 Man's View

So, God entangled the dark with the light:

$$\langle \sqrt{D}\sqrt{D+L}, \sqrt{L}\sqrt{D+L} \rangle$$

If

$$\int \sqrt{f}\, d\omega = -5.669067105 \times 10^5 + i53102.41304,$$

then let

$$\sqrt{x} = -5.669067105 \times 10^5, \qquad \sqrt{y} = 53102.41304$$

And

$$\langle \sqrt{x}, \sqrt{y} \rangle \to \langle \sqrt{x}\sqrt{x+y}, \sqrt{y}\sqrt{x+y} \rangle$$
$$= -3.227900723 \times 10^{11} + i3.023589495 \times 10^{10}$$

So, God entangled the speed of the dark with the speed of the light:

$$Speed\ of\ the\ mostly\ dark = 3.227900723 \times 10^{11},$$

$$Speed\ of\ the\ mostly\ light = 3.023589495 \times 10^{10},$$

which is not their actual speeds. For God forbids Man from knowing their actual speeds because he ate of the forbidden fruit. Out of this, God created "space-time." Space-time is an entanglement between space and time, but they are perfect and separate in Eden. But Man cannot see Eden.

Now God needed something out of which to make the parts of Man's body. But Satan is embodied in Man, and therefore, he will know death. God made Man's body out of "charge" and "mass," and that became Man's body:

$$-\int_0^a \sqrt{g}\, d\omega = 1.622335843 \times 10^{-19} + i7.956265238 \times 10^{-31},$$

where $a = 3.111669925 \times 10^{-12}$ is Satan's number. The first number is "charge," and the second number "mass." To show that God is part of Man's body, factor out His number:

$$-\int_0^a \sqrt{g}\, d\omega = \sqrt{G}(3.111669925 \times 10^{-12} + i1.526026276 \times 10^{-23})$$

The number "$3.111669925 \times 10^{-12}$" is Satan's number, which is embodied in Man so that he will die. Physicists call the second number the "Higgs boson." It provides all things with "mass."

Now if

$$\sqrt{x} = 1.622335843 \times 10^{-19}, \qquad \sqrt{y} = 7.956265238 \times 10^{-31},$$

then God takes

$$\langle \sqrt{x}, \sqrt{y} \rangle \to \langle \sqrt{x}\sqrt{x+y}, \sqrt{y}\sqrt{x+y} \rangle$$
$$= 2.631973587 \times 10^{-38} + i1.29077342710^{-49},$$

which physicists call the "atom." The atom is an entanglement between mass and charge. Today, physicists envision the atom as two equal but opposite charges—the "proton" and the "electron" with their masses. But this is not so. The proton and the electron are not separate, lest Man could see Eden. So, Man's body consists of atoms.

3.2.4 Man's Place

God drove Man from Eden but had to make a place for him to live. It became a big place, but Man would be small in it, and he will be alone and will never leave his place. He will be alone all the days of his life. And Man will rule over the place where God put him, but he will be alone." So, God made Man's place a big place but put Man in a small place inside. And here's how God used His function and His number:

$$\int \sqrt{f'}\, d\omega = -1.087337476 \times 10^{13} + i1.018514029 \times 10^{12},$$

$$\sqrt{f'} = \left(-\frac{1}{\sqrt{G}}\right)^{1-\omega} (\sqrt{G})^{\omega-1}$$

So, the big place God made for Man to live has both dark-parts and light-parts. And He separated the dark from the light. The place that God made for Man is big. And it will become bigger and bigger all the days of its life. God entangled the light-parts with the dark-parts, so Man could not know what He had made. So, Man cannot see Eden. The square root of the dark-part "$\sqrt{x}$" is:

$$\sqrt{x} = -1.087337476 \times 10^{13},$$

The square root of the light-part "$\sqrt{y}$" is:

$$\sqrt{y} = 1.018514029 \times 10^{12}$$

God entangling them is this:

$$\langle \sqrt{x}\sqrt{x+y}, \sqrt{y}\sqrt{x+y} \rangle = -1.187478315 \times 10^{26} + i1.112316415 \times 10^{25}$$

Now if

$$\bar{v}_D^2 = (-1.187478313 \times 10^{26})^2, \qquad -\bar{v}_L^2 = -(1.112316413 \times 10^{25})^2$$

then,

$$\sqrt{\bar{v}_D^2 - \bar{v}_L^2} = 1.182257276 \times 10^{26}$$

So,

$$\frac{1}{\sqrt{\bar{v}_D^2 - \bar{v}_L^2}} = 8.458395819 \times 10^{-27}$$

Today, physicists call the number "$8.458395819 \times 10^{-27}$" "Planck's constant." But it is not a constant of proportionality. It is a charge. Man did not know this. But he knows it now.

God made the heavens, both the dark and the light. So Man could see with his eyes the things that God made, He took "g" and made the stuff of light:

$$-\int_0^a \sqrt{g'}\, d\omega = 8.458395814 \times 10^{-27} + i4.148169497 \times 10^{-38},$$

$$\sqrt{g'} = \left(-\frac{1}{\sqrt{G}}\right)^{\omega-1} \left(\sqrt{G}\right)^{-\omega+1}$$

The dark-part is called "charge," and the light-part is called "mass." The dark-part is "Planck's constant." And the mass comes from the Higgs. The light-part today is called the "electron neutrino." Man does not know that God made these, but I will prove it to you:

$$-\int_0^a \sqrt{g'}\, d\omega = G(3.111669925 \times 10^{-12} + i1.526026276 \times 10^{-23}),$$

$$a = 3.111669925 \times 10^{-12}$$

The number "$3.111669925 \times 10^{-12}$" is Satan's number. And the number "$1.526026276 \times 10^{-23}$" is the "Higgs" because it is mass. And "G" is God's number.

So, God made charge and mass to make the light so Man could see with his eyes what God had made:

$$\sqrt{x} = 8.458395814 \times 10^{-27}, \quad \sqrt{y} = 4.148169497 \times 10^{-38}$$

God entangled the mass with the charge because Man had disobeyed:

$$\langle \sqrt{x}, \sqrt{y} \rangle \rightarrow \sqrt{x}\sqrt{x+y} + i\sqrt{y}\sqrt{x+y}$$
$$\approx 7.154445975 \times 10^{-53} + i3.508685951 \times 10^{-64}$$

Hence, the light particle is made of mass and charge. It is both dark and light. But its mass and charge are small and entangled, so Man did not know it is a light particle. But now he knows.

So, God created only two things out of which the substance of the universe consists. The first, Man calls the "atom." The second, Man calls the "photon" (see Table 3.2.4-1).

Standard Particles	
$\sqrt{g(t)}$	$\sqrt{g_-(t)}$
+	−
Atom	Atom
Photon	Photon

Table 3.2.4-1

3.2.5 Satan

Now God created Lucifer, but Lucifer fell to his vanity and became Satan when seducing the woman. For that, God made him small. He is the "Prince of Darkness" because he is finite. So, Satan's number is:

$$3.111669925 \times 10^{-12}$$

From where does Satan's number come? God entangled the light-speed with the dark-speed:

$$Speed\ of\ the\ mostly\ dark = 3.227900723 \times 10^{11},$$

$$Speed\ of\ the\ mostly\ light = 3.023589495 \times 10^{10}$$

If

$$\bar{v}_D^2 = (3.227900723 \times 10^{11})^2, \quad -\bar{v}_L^2 = -(3.023589495 \times 10^{10})^2$$

then,

$$\sqrt{\bar{v}_D^2 - \bar{v}_L^2} = 3.213708473 \times 10^{11}$$

So,

$$\frac{1}{\sqrt{\bar{v}_D^2 - \bar{v}_L^2}} = 3.111669925 \times 10^{-12}$$

And this is where Satan's number comes from. But Satan pervades all that God has made. Satan is small but everywhere.

Now Satan is angry because God made him small. Satan is finite. So, if he could make God small, he would rule over God. Thus, Satan will deceive Man by claiming that $G = 0$ so that he can make himself higher than God. Satan is cunning and will lead many astray. But if $G = 0$, f and g are destroyed. So, Satan's way leads to destruction.

3.3 Concluding Remarks

1-1.1 In the beginning God created the heavens and the earth. 2 And the earth was waste and void; and darkness was upon the face of the deep: and the Spirit of God moved upon the face of the waters. 3 And God said, "Let there be light." And there was light. 4 And God saw the light, that it was good: and God divided the light from the darkness.

And with this knowledge, Man can return to God.

Chapter 4

How God Made Space-Time

"1-1.16 God made the two great lights—the greater light to rule the day and the lesser light to rule the night—and the stars."

— *The First Book of Moses*

God made a place for Man, which he calls "space." And God saw that it was good that Man should have a body that changed so that a man could live his life. The change was called "time." So, God created space and time so that Man could live his life, but then Man would die.

So, God created space "S" and time "T." So that Man could not tell the space from the time, God entangled them into space-time using His complex numbers:

$$\langle T, S \rangle \xrightarrow{f} \langle \sqrt{T}\sqrt{S+T}, \sqrt{S}\sqrt{S+T} \rangle$$

God had driven Man from Eden, lest he would take from the tree of life and live forever. But, God had more than one way of representing space-time:

$$\langle T, S \rangle \xrightarrow{-if} \langle \sqrt{S}\sqrt{S+T}, -\sqrt{T}\sqrt{S+T} \rangle$$

4.1 The Principle of Relativity

Now God made both the dark and the light. But God entangled the dark with the light, so Man could not be sure which was which because Man had disobeyed God. If the maximum mostly light-speed is "c" and the maximum primarily dark-speed is "c'," then

$$\left\langle \left(\frac{d\bar{S}}{dt}\right)_D, \left(\frac{d\bar{S}}{dt}\right)_L \right\rangle = \langle -c', c \rangle,$$

where $(d\bar{S}/dt)_D$ is the maximum mostly dark-speed and $(d\bar{S}/dt)_L$ is the maximum mostly light-speed. But another way is:

$$\left|\left(\frac{d\bar{S}'}{dt'}\right)_L - \left(\frac{d\bar{S}'}{dt'}\right)_D\right| = \langle c, c'\rangle$$

(see Fig. 4.1-1 a), b)). Now the algebra:

$$-\left(\frac{d\bar{S}}{dt}\right)_L^2 = -c^2, \quad \left(\frac{d\bar{S}'}{dt'}\right)_L^2 = c^2 \rightarrow -d\bar{S}_L^2 = -c^2\, dt^2, \quad (d\bar{S}_L')^2 = (c\, dt')^2$$

Adding the last two equations on the right above leaves

$$(d\bar{S}_L')^2 - d\bar{S}_L^2 = (c\, dt')^2 - c^2\, dt^2 \rightarrow d\bar{S}_L^2 - c^2\, dt^2 = (d\bar{S}_L')^2 - (c\, dt')^2$$

The last equation on the right above expands to three spatial dimensions by setting

$$d\bar{S}_L^2 = dx_L^2 + dy_L^2 + dz_L^2, \quad (d\bar{S}_L')^2 = (dx_L')^2 + (dy_L')^2 + (dz_L')^2$$

The result is

$$dx_L^2 + dy_L^2 + dz_L^2 - c^2\, dt^2 = (dx_L')^2 + (dy_L')^2 + (dz_L')^2 - (c\, dt')^2,$$

which is the "principle of relativity."

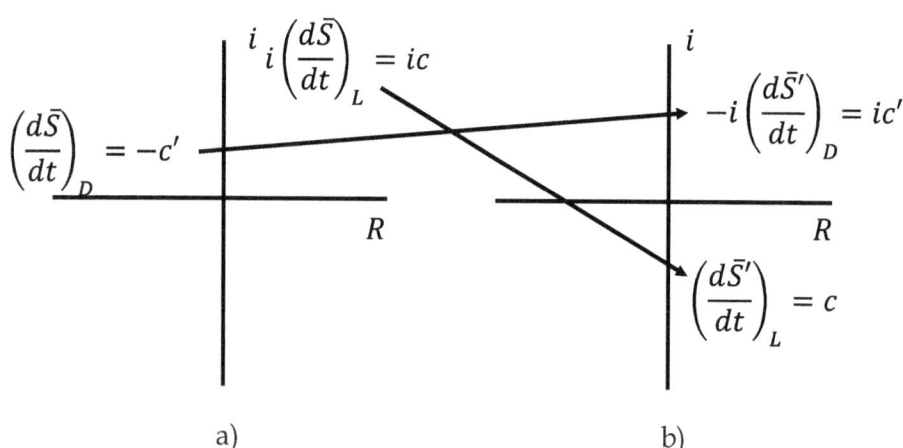

a) b)

Figure 4.1-1

There is a mostly dark "relativity principle" identical to the one for the mostly light, i.e.,

$$dx_D^2 + dy_D^2 + dz_D^2 - (c'dt)^2 = (dx_D')^2 + (dy_D')^2 + (dz_D')^2 - (c'dt')^2$$

God made space-time so that space and time became one, and Man could not tell them apart. God took space and entangled it with time, so Man is confused and cannot see God.

4.1.2 Lorentz Transformations

The equations below are called the "Lorentz transformations," but Man only knows the light-part equations. He did not know them for the dark-part. Now he does.

Light			
Position		Time	
$x' = \dfrac{x - vt}{\sqrt{1 - \dfrac{v^2}{c^2}}}$,	$x = \dfrac{x' + vt'}{\sqrt{1 - \dfrac{v^2}{c^2}}}$	$t' = \dfrac{t - \dfrac{v}{c^2}x}{\sqrt{1 - \dfrac{v^2}{c^2}}}$,	$t = \dfrac{t' + \dfrac{v}{c^2}x'}{\sqrt{1 - \dfrac{v^2}{c^2}}}$

Dark			
Position		Time	
$x' = \dfrac{x - \dfrac{v}{k}t}{\sqrt{1 - \dfrac{v^2}{k^2c^2}}}$,	$x = \dfrac{x' + \dfrac{v}{k}t'}{\sqrt{1 - \dfrac{v^2}{k^2c^2}}}$	$t' = \dfrac{t - \dfrac{v}{kc^2}x}{\sqrt{1 - \dfrac{v^2}{k^2c^2}}}$,	$t = \dfrac{t' + \dfrac{v}{kc^2}x'}{\sqrt{1 - \dfrac{v^2}{k^2c^2}}}$
$k = \dfrac{c'}{c}$		$k = \dfrac{c'}{c}$	

4.2 Time

Once God created the physics of the universe, He created biology, which is based on physics. 1-1.7 And God made the firmament, and divided the waters which were under the firmament from the waters which were above the firmament: and it was so.

1-1.9 And God said, "Let the waters under the heavens be gathered together unto one place, and let the dry land appear." And it was so. 10 And God called the dry land earth; and the gathering together of the waters He called Seas: and God saw that it was good. 11 And God said, "Let the earth put forth grass, herbs yielding seed, and fruit trees bearing fruit after their kind, wherein is the seed thereof, upon the earth." And it was so. 12 And the earth brought forth grass, herbs yielding seed after their kind, and trees bearing fruit, wherein is the seed thereof, after their kind: and God saw that it was good.

1-1.20 And God said, "Let the waters swarm with living creatures, and let birds fly above the earth in the open firmament of Heaven." 21 And God created the great sea-monsters, and every living creature that moveth, wherewith the waters swarmed, after their kind, and every winged bird after its kind: and God saw that it was good. 22 And God blessed them, saying, "Be fruitful, and multiply, and fill the waters in the seas, and let birds multiply on the earth."

So that the creatures brought forth by God could live, God created "time." And all that God created would grow proportional to His time. So, Man's body would grow proportional to God's time. And this is how God used His number to create His time:

$$\bar{v}(t) = Gt,$$

where $\bar{v}(t)$ is the rate of God's time relative to the rate of Man's time. Man can relate God's time to Man's time because Man's time is artificial, so he does not know "time." Man's time related to God's time looks more like this:

$$\bar{t} = \frac{t}{\sqrt{1 - \frac{\bar{v}(t)^2}{c^2}}} \rightarrow \bar{t} = \frac{t}{\sqrt{1 - \frac{G^2 t^2}{c^2}}},$$

where t is "Man's time" and is arbitrary, and $\bar{t}$ is "God's time" relative to "Man's time." But since Man does not know "time," he sees time as space growing, space expanding, like his body does, because he does not know "time." The equation that relates distance to time is this:

$$\int \frac{dx}{\sqrt{1 - \frac{[l]^2 G^2 x^2}{c^4}}} = c \int dt,$$

where "x" is a distance, "t" is time, and "l" represents the dimensions that make the equation consistent. Of course, "c" is the mostly light-speed, and "G" is God's number.

In Man's time, it would take about 1/3 of a second for the universe to expand to a diameter of approximately 10^{10} [cm], which is about 62,137 [miles]. That is about 7% of the diameter of the Sun or about 7.7 Earth diameters. Expanding the universe to a size of approximately 10^{20} [cm] across, which is about 6.2×10^{14} [miles] or about 106 [light − years] across, would take approximately 106 [years]. It would take about 10^8 [billion years] for the universe to expand to a diameter of ≈ 10^{17} [light − years].

Man's equation for how things move in space-time:

$$G_{\mu\nu} + \Lambda g_{\mu\nu} = -\frac{8\pi\sqrt{G}}{c^4} T_{\mu\nu}$$

It does not tell Man whether the universe expands, contracts, or is static. It only says how things and light particles move through space-time due to gravity. But it is barren because God's time is absent in it. Man's space-time lets the universe be flat, but God does not allow the universe nor the Earth to be flat. The universe will expand forever. It is not static. The new equation should look more like this:

$$R_{\mu\nu} = -8\pi\gamma T_{\mu\nu}, \quad \gamma = \frac{3\pi t - \cos(\pi t)\sin(\pi t) + 2\pi}{2\pi}, \quad G, c = 1,$$

which includes God's time.

The symbol "Λ" in Man's equation for gravity is called the "cosmological constant." It is not needed because of God's time:

$$\Lambda \approx 1.2 \times 10^{-50}$$

$$God's\ time\ proportion = \frac{G^2}{c^2} \approx 8.21 \times 10^{-51}$$

Note that

$$\Lambda \approx \frac{G^2}{c^2}$$

4.3 Concluding Remarks

1-2.1 And the heavens and the earth were finished, and all the host of them. 2 And on the seventh day God finished all that He had made; and He rested on the seventh day from all that He had made. 3 And God blessed the seventh day, and hallowed it; because that in it He rested from all his work … 4 These are the generations of the heavens and of the earth when they were created, in the day that God made earth and Heaven.

Chapter 5

Epilogue

The Nature of the Universe

"No one has seen God at any time; if we love one another, God abides in us, and His love is perfected in us."

- 1 John 4:12

https://bible.knowing-jesus.com/topics/Not-Seeing-

It is time for Man to return to God — to Eden, the place from which he came because God has decided that Man has suffered enough. And God will point the way.

5.1 Small-Scale Physics

The form of "g," the physics on the small-scale:

$$g(\theta) = \frac{-1}{\sqrt{e}} \left(\frac{-1}{\sqrt{e}}\right)^{-1+\theta} \left(\sqrt{e}\right)^{-\theta} = e^{(i\pi-1)\theta} = e^{-\theta}(\cos(\pi\theta) + i\sin(\pi\theta))$$

5.2 Large-Scale Physics

The form of "f," the physics on the large-scale[3]:

$$f(\theta) = -\sqrt{e}\left(\frac{-1}{\sqrt{e}}\right)^{1-\theta} \left(\sqrt{e}\right)^{\theta} = e^{(1-i\pi)\theta} = e^{\theta}(\cos(\pi\theta) - i\sin(\pi\theta))$$

---

[3] Proofs of these relations appear in Appendix "A2."

5.3 Combining the Small with the Large

Now

$$e^{(1-i\pi)\theta} = e^{\theta}(\cos(\pi\theta) - i\sin(\pi\theta))$$

is a complex spiral. The square root of $e^{(1-i\pi)\theta}$ is

$$e^{1/2(1-i\pi)\theta} = e^{\theta/2}\left(\cos\left(\frac{\pi}{2}\theta\right) - i\sin\left(\frac{\pi}{2}\theta\right)\right)$$

But God entangles the real with the imaginary part of the complex number:

$$\overline{e^{(1-i\pi)\theta}} = \left\langle e^{\theta/2}\cos\left(\frac{\pi}{2}\theta\right), -e^{\theta/2}\sin\left(\frac{\pi}{2}\theta\right)\right\rangle \sqrt{e^{\theta}\cos^2\left(\frac{\pi}{2}\theta\right) + e^{\theta}\sin^2\left(\frac{\pi}{2}\theta\right)}$$
$$= e^{\theta}\left\langle \cos\left(\frac{\pi}{2}\theta\right), -\sin\left(\frac{\pi}{2}\theta\right)\right\rangle = e^{\theta}\left(\cos\left(\frac{\pi}{2}\theta\right) - i\sin\left(\frac{\pi}{2}\theta\right)\right),$$
$$\cos^2\left(\frac{\pi}{2}\theta\right) + \sin^2\left(\frac{\pi}{2}\theta\right) = 1$$

Now $e^{(1-i\pi)\theta}$ is a complex spiral with period "2." And $\overline{e^{(1-i\pi)\theta}}$ is the same complex spiral, but with period "4" (see Fig. 5.3-1 a), b)).

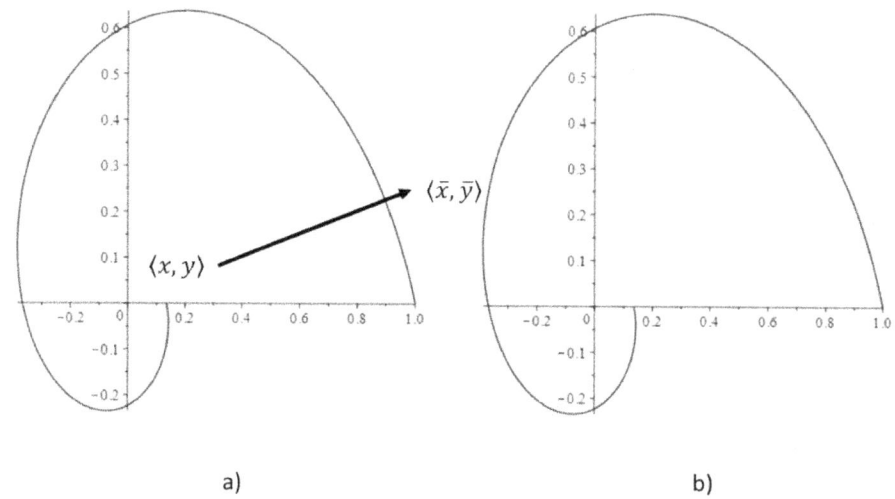

a)   b)

Figure 5.3-1

EPILOGUE

On the other hand, on the small-scale:

$$e^{-(1-i\pi)\theta} = e^{-\theta}(\cos(\pi\theta) + i\sin(\pi\theta))$$

is also a complex spiral with period "2." The square root of $e^{-(1-i\pi)\theta}$ is

$$e^{-1/2(1-i\pi)\theta} = e^{-\theta/2}\left(\cos\left(\frac{\pi}{2}\theta\right) + i\sin\left(\frac{\pi}{2}\theta\right)\right)$$

But God entangles the real with the imaginary part of the complex number:

$$\overline{e^{-(1-i\pi)\theta}}$$
$$= \left\langle e^{-\theta/2}\cos\left(\frac{\pi}{2}\theta\right), e^{-\theta/2}\sin\left(\frac{\pi}{2}\theta\right)\right\rangle \sqrt{e^{-\theta}\cos^2\left(\frac{\pi}{2}\theta\right) + e^{-\theta}\sin^2\left(\frac{\pi}{2}\theta\right)}$$
$$= e^{-\theta}\left\langle \cos\left(\frac{\pi}{2}\theta\right), \sin\left(\pi\frac{\pi}{2}\theta\right)\right\rangle = e^{-\theta}\left(\cos\left(\frac{\pi}{2}\theta\right) + i\sin\left(\frac{\pi}{2}\theta\right)\right),$$
$$\cos^2\left(\frac{\pi}{2}\theta\right) + \sin^2\left(\frac{\pi}{2}\theta\right) = 1$$

The function "$e^{-(1-i\pi)\theta}$" is a complex spiral with period "2." And $\overline{e^{-(1-i\pi)\theta}}$ is the same complex spiral, but with period "4" (see Fig. 5.3-2 a), b)).

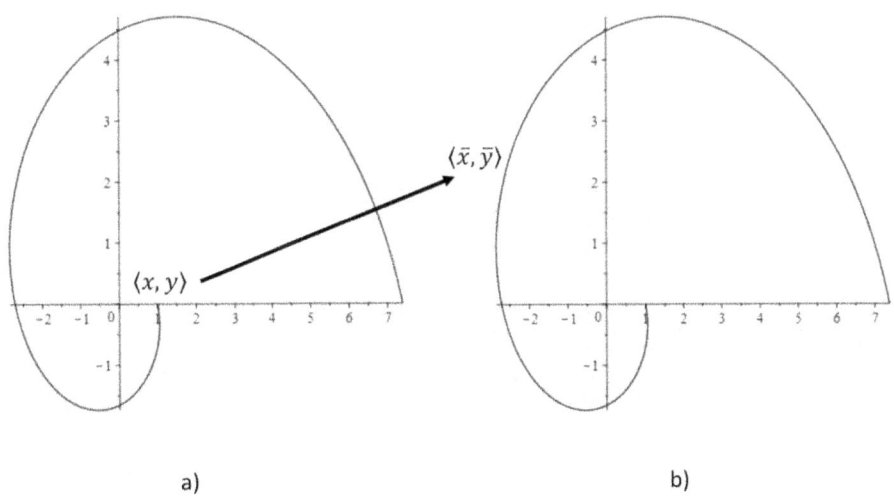

a) b)

Figure 5.3-2

5.4 The Wave Function

The wave function for life is

$$\frac{d\tau}{dt} = 1 + \sin^2(\pi t) \rightarrow \frac{1}{\left(\frac{d\tau}{dt}\right)} = \frac{1}{1+\sin^2(\pi t)}, \quad \frac{d\tau}{dt}\frac{1}{\left(\frac{d\tau}{dt}\right)} = 1, \quad G,c = 1,$$

one equation for the large-scale, the other for the small-scale (see Fig. 5.4-1). God's time "τ" is called "cosmic time," while "t" is "Man's time." The change in God's time relative to the change in Man's time looks like this:

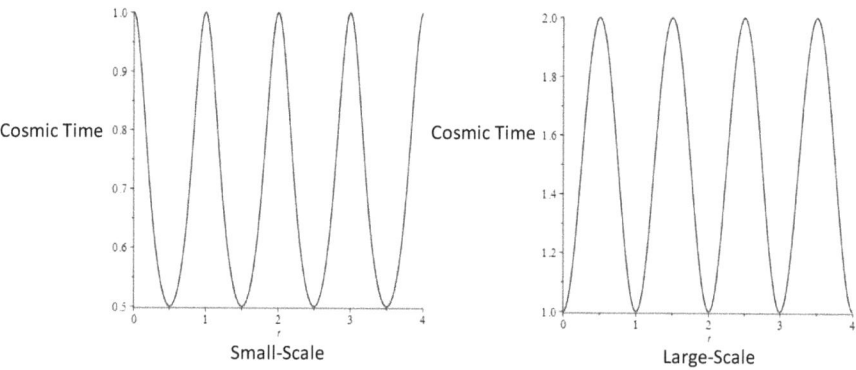

Figure 5.4-1

The change in the energy with spacial expansion "dE/dx" for life if $E = \tau$ and $t = x$ more to scale (Fig. 5.4-2):

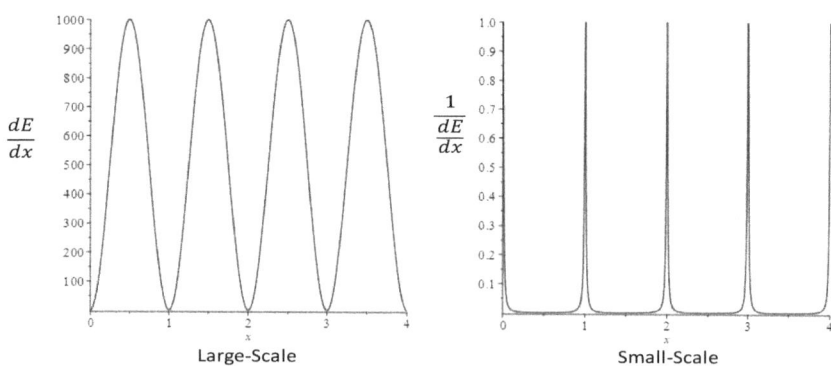

Figure 5.4-2

Hence,

$$E = \int 1 + \sin^2(\pi x)\, dx \to E = \frac{3\pi x - \cos(\pi x)\sin(\pi x) + 2\pi}{2\pi},$$

(see Fig. 5.4-3).

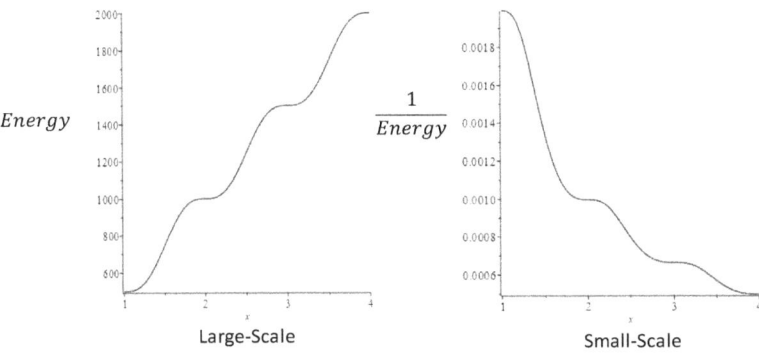

Figure 5.4-3

5.5 Final Remarks

Such is the cycle of life. The change in energy comes to the large physics from the small physics, but then the change in energy reverses. So, a man's life, borrowed from the dust, returns to it, which physicists call the "conservation of energy." Today, Man fears the freedom he gained by partaking of the fruit of the tree of the knowledge of good and evil. He fears the freedom to choose his path and wants to return to the womb, to Eden, where he was born and is safe with God. Man now craves for God to save him from himself by sending His Son once more.

But, God gave Man "free will" and the right to choose, so Man ate the fruit of the knowledge of good and evil of his own volition. Because Man disobeyed God's word, God blocked the way back to Eden. Man hates God and those like himself because God made Man in His likeness and let Man choose.

Man's destiny has but two outcomes. Man can fear his freedom and return to the dust of the ground, or overcome his fear and return to Eden by the power of the mind that God gave Him. It is now time for Man to return to God, to Eden, his home. But Man must choose the path to Eden over the path to dust by overcoming His fear of freedom and replacing it with a love of Mankind. God Bless.

Appendices

A1.

Operations on Complex Numbers[4]

1. Addition:

$$\langle a, b \rangle + \langle c, d \rangle = \langle a+c, b+d \rangle$$
$$\xrightarrow{f} \left\langle \left(\sqrt{(a+c)'}\right), \left(\sqrt{(b+d)'}\right) \right\rangle \sqrt{\left(\sqrt{(a+c)'}\right)^2 + \left(\sqrt{(b+d)'}\right)^2}$$
$$= \left\langle \sqrt{(a+c)'}, \sqrt{(b+d)'} \right\rangle \sqrt{|a+c| + |b+d|}$$

2. Multiplication:

$$\langle a, b \rangle \cdot \langle c, d \rangle$$
$$= \langle ac - bd, ad + bc \rangle$$
$$\xrightarrow{f} \left\langle \sqrt{(ac-bd)'}, \sqrt{(ad+bc)'} \right\rangle \sqrt{\left(\sqrt{(ac-bd)'}\right)^2 + \left(\sqrt{(ad+bc)'}\right)^2}$$
$$= \left\langle \sqrt{(ac-bd)'}, \sqrt{(ad+bc)'} \right\rangle \sqrt{|ac-bd| + |ad+bc|}$$

3. Additive Identity:

$$\langle 0,0 \rangle \xrightarrow{f} \langle \sqrt{0}, \sqrt{0} \rangle \sqrt{0} = \langle 0,0 \rangle$$

4. Additive Inverse:

$$\langle -a, -b \rangle \xrightarrow{f} \left\langle -\sqrt{a'}, -\sqrt{b'} \right\rangle \sqrt{|a| + |b|}$$

---

[4] The rules are valid if and only if

$$\langle a, b \rangle \to \sqrt{a'} = \begin{cases} \sqrt{a}, & \text{if } a \geq 0 \\ -\sqrt{|a|}, & \text{if } a < 0 \end{cases}, \quad \sqrt{b'} = \begin{cases} \sqrt{b}, & \text{if } b \geq 0 \\ -\sqrt{|b|}, & \text{if } b < 0 \end{cases}, \quad a, b \in R$$

5. Multiplicative Identity:

$$\langle 1,0 \rangle \xrightarrow{f} \langle \sqrt{1}, \sqrt{0} \rangle \sqrt{1} = \langle 1,0 \rangle$$

6. Multiplicative Inverse:

$$\left\langle \frac{a}{a^2+b^2}, \frac{-b}{a^2+b^2} \right\rangle \xrightarrow{f} \left\langle \frac{\sqrt{a'}}{(|a|+|b|)^2}, \frac{-\sqrt{b'}}{(|a|+|b|)^2} \right\rangle \sqrt{|a|+|b|}, \qquad a, b \neq 0$$

Note that $i^2 = -1$.

A2.

Consider

$$g(x) = \left(\frac{-1}{\sqrt{e}} \right)^{-1+x} (\sqrt{e})^{-x}$$

Expanding $g(x)$ in a power series gives

$$g(x) = \left(\frac{-1}{\sqrt{e}} \right)^{-1+x} (\sqrt{e})^{-x}$$

$$= -\sqrt{e} - \sqrt{e}(-1 + i\pi)x + \frac{\sqrt{e}(-1 + 2\pi + \pi^2)x^2}{2!}$$

$$+ \frac{\sqrt{e}(1 - 3i\pi - 3\pi^2 + i\pi^3)x^3}{3!}$$

$$- \frac{\sqrt{e}(1 - 4i\pi - 6\pi^2 = 4i\pi^3 + \pi^4)x^4}{4!}$$

$$- \frac{\sqrt{e}(-1 + 5i\pi + 10\pi^2 - 10i\pi^3 - 5\pi^4 + i\pi^5)x^5}{5!} + \cdots,$$

which can be written

$$g(x) = \left(\frac{-1}{\sqrt{e}}\right)^{-1+x} (\sqrt{e})^{-x}$$
$$= -\sqrt{e} - \sqrt{e}(-1+i\pi)x - \frac{\sqrt{e}(-1+i\pi)^2 x^2}{2!} - \frac{\sqrt{e}(-1+i\pi)^3 x^3}{3!}$$
$$- \frac{\sqrt{e}(-1+i\pi)^4 x^4}{4!} - \frac{\sqrt{e}(-1+i\pi)^5 x^5}{5!} + \cdots$$

Let
$$y = (-1+i\pi)x,$$
then
$$g(x) = \left(\frac{-1}{\sqrt{e}}\right)^{-1+x} (\sqrt{e})^{-x} = -\sqrt{e}\left(1 + y + \frac{y^2}{2!} + \frac{y^3}{3!} + \frac{y^4}{4!} + \cdots\right) = -\sqrt{e}\, e^y$$
$$= -\sqrt{e}\, e^{(-1+i\pi)x} \to \left(\frac{-1}{\sqrt{e}}\right)^{-1+x} (\sqrt{e})^{-x} = -\sqrt{e}\, e^{-x} e^{i\pi x}$$
$$= -\sqrt{e}(\sqrt{e})^{-x}(\sqrt{e})^{-x} e^{i\pi x} \to \left(\frac{-1}{\sqrt{e}}\right)^x = (\sqrt{e})^{-x} e^{i\pi x}$$
$$\to \left(\frac{-1}{\sqrt{e}}\right)^x (\sqrt{e})^x = (-1)^x = e^{i\pi x} = \cos(\pi x) + i\sin(\pi x)$$

Moreover,
$$\frac{-1}{\sqrt{e}}\left(\frac{-1}{\sqrt{e}}\right)^{-1+x} (\sqrt{e})^{-x} = e^{-x} e^{i\pi x} = e^{(i\pi-1)x}$$

List of References

12. D. Tong *(2009)*, "*String Theory,*" University of Cambridge Part III Mathematical Tripos, Department of Applied Mathematics and Theoretical Physics, Centre for Mathematical Sciences, Wilberforce Road, Cambridge, CB3 OWA, UK.

13. P. Woit *(2006)*, "*Not Even Wrong: The Failure of String Theory and the Search for Unity in Physical Law,*" Perseus Books, NY.

14. L. Smolin *(2007)*, "*The Trouble with Physics: The Rise of String Theory: The Fall of Science and What Comes Next,*" Houghton Mifflin Co., Boston.

15. L. Susskind *(2003)*, "*The Anthropic Landscape of String Theory,*" Department of Physics Stanford University, Stanford, CA 94305-4060.

16. A. Z. Jones with D. Robbins *(2010)*, "*String Theory for Dummies,*" Wiley Publishing, Inc., NJ.

17. D. Gross *(2005)*, "*Einstein and the Search for Unification,*" Current Science, Vol. 89, No. 12, 25 December 2005.

37. E. Witten *(2002)*, "Can scientists' 'theory of everything' really explain all the weirdness the universe displays?" www.astronomy.com.

179. S. Lipschutz *(1964)*, "*The Theory and Problems of Set Theory and Related Topics,*" Schaum's Outlines Series, McGraw-Hill.

281. The Holy Bible *(1946)*, "The First Book of Moses," Thomas Nelson & Sons.

282. T. Coburn *(2019)*, "Image Theory," www.roadtoatoe.net.

283. E. A. Nilsen, T. Coburn *(2020)*, "The Fractal Acoustoelectric Origin of Mass, Gravity, and Light," Draft.

www.ingramcontent.com/pod-product-compliance
Lightning Source LLC
Chambersburg PA
CBHW061605110426
42742CB00039B/2854